Mathematics to Sixteen
Book 3

by

R. L. BOLT, M.Sc.

Senior Mathematics Master
Woodhouse Grove School

and

C. REYNOLDS, M.Sc.

Lecturer,
School of Education
University of Leeds

UNIVERSITY TUTORIAL PRESS

Published by University Tutorial Press Ltd
842 Yeovil Road, Slough SL1 4JQ

Published 1978

ISBN: with Answers: 0 7231 0745 9
without Answers: 0 7231 0754 8

Printed in Great Britain by Page Bros (Norwich) Ltd, Norwich

CONTENTS

Contents

PREFACE

This is the third of four books providing a modern course in Mathematics for secondary pupils of average ability. The course covers the syllabuses of the present CSE examinations and much of the work required for the GCE 'O' Level examinations and so will be suitable for most examinations for pupils of 16 years of age.

This book introduces geometric transformations, trigonometric ratios, topology, vectors and probability whilst continuing to develop the arithmetic, algebra and statistics of the other books.

Here, as in Books 1 and 2, topics are approached in a variety of ways. It is expected that some, such as the algebra sections, will be taught to classes as a whole and in these sections the text is brief: in some other sections, such as the one on the tangent ratio, a preliminary exercise introduces new ideas or enables pupils to discover and use new facts which can then be consolidated by the teacher: other sections, for example those on transformations and topology, can be explored by pupils with very little assistance from the teacher.

In the two earlier books, the questions were designed to be done without any calculating aids: in this book aids are needed. There are sections on the use of four figure tables of squares, square roots, logarithms and trigonometric ratios. For these sections pupils will require a separate book of tables. As electronic calculators are now used by many pupils and are permitted by many examining boards, we have also included a section on them. Although this section is placed at the end of the book, it can be taken much earlier in the course if desired. In it we emphasise the importance of examining answers to see whether they are reasonable, of giving answers to a sensible number of figures and of not using calculators for very simple calculations.

To do Mathematics successfully and to enjoy doing it, certain skills and techniques are essential. To develop such skills and techniques, and indeed to understand and appreciate many of the concepts and applications of Mathematics, a certain amount of repetitive practice is needed. We hope that teachers will find sufficient practice material in this book. There are three sets of Revision Papers which include questions on the work of Books 1 and 2 as well as on the work of this book.

R L Bolt
C Reynolds

1 · TRANSFORMATIONS

In moving to a new school or a new house we are particularly interested in those things that are different and in those that remain the same. When objects or shapes undergo movements or translations we like to know what remains constant and what is changed.

TRANSLATIONS

Place a set square, ABC, against a ruler on your desk as shown in Fig. 1. Slide the set square 15 cm along the edge of the ruler to position $A_1B_1C_1$. The set square has moved 15 cm in a straight line in the direction shown by the arrow.

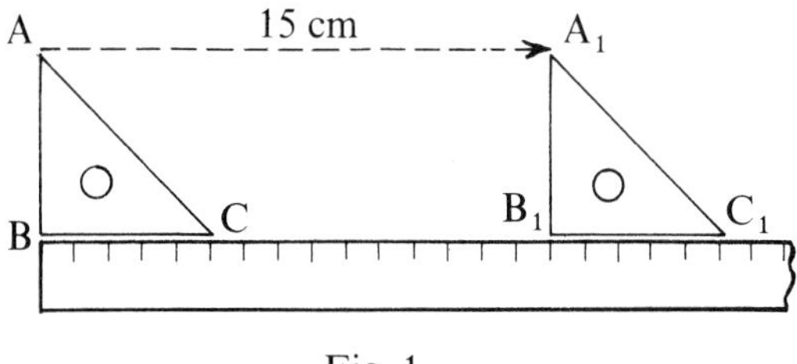

Fig. 1

Point A_1 is called the *image* of point A.

Triangle $A_1B_1C_1$ is called the image of triangle ABC.

Such a movement is called a *translation*. An object undergoes a translation when it moves a fixed distance in a given direction. The movement takes place in a straight line without turning. The translation in this case is represented by this arrow.

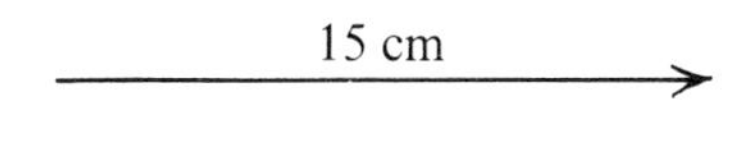

Exercise 1

1. Use a 45° set square or a triangle of the same shape made of card. Place the triangle ABC on a square grid and let side AB be 1 unit. Draw round $\triangle$ABC. Move it 2 units upwards.

Draw round it again and label the new position $A_1B_1C_1$.
What do you notice about the lengths of the lines AA_1, BB_1 and CC_1?

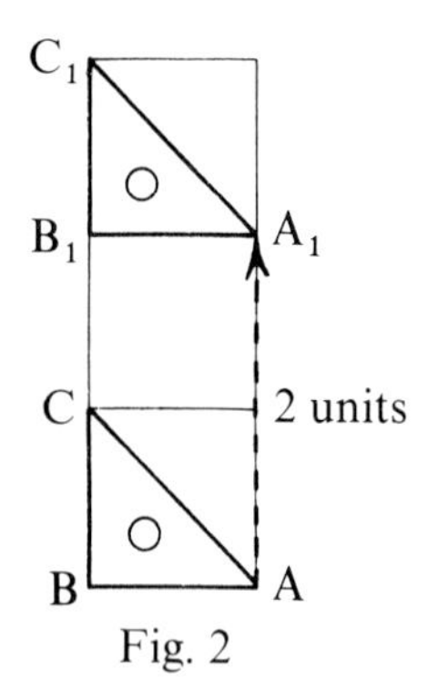

Fig. 2

2. Use the same triangle ABC. Place it as before and move it to occupy the position $A_1B_1C_1$ shown in Fig. 3. Draw and measure the lengths of the lines AA_1, BB_1 and CC_1. Are they equal in length? Are they parallel?

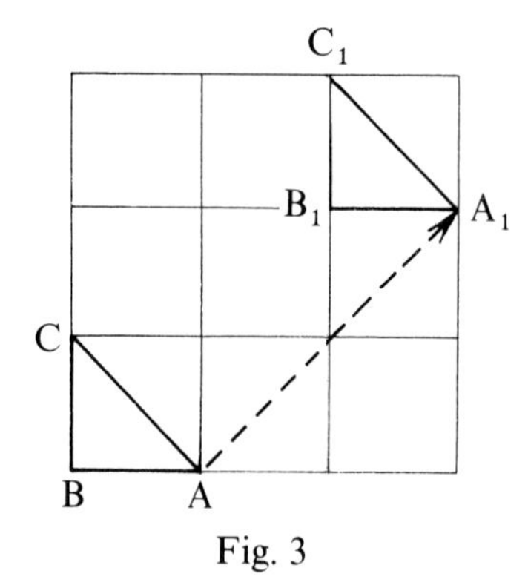

Fig. 3

Represent the translation by a suitable arrow.

3. In Questions **1** and **2**:

(i) What is the image of AB and AC?
(ii) Is A_1B_1 equal and parallel to AB?
(iii) Is A_1C_1 equal and parallel to AC?

4. Copy the shapes of Fig. 4 onto squared paper. Apply the translation shown by each arrow and draw the image of each shape. In each case point A_1 is the image of point A.

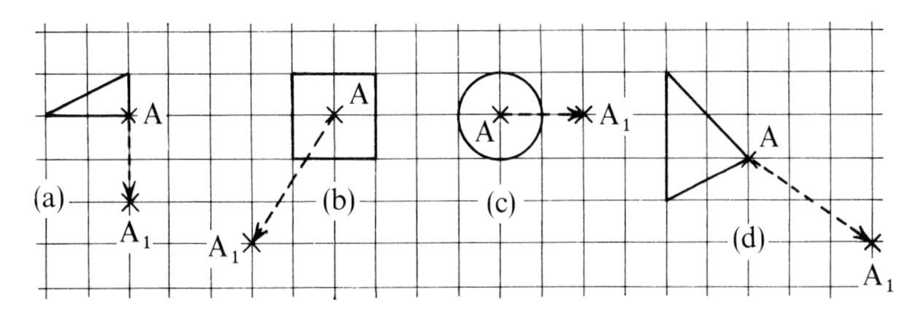

Fig. 4

5. Fig. 5 shows the images of shapes that have undergone the translations shown by the arrows. Copy the diagrams and draw the original positions of the shapes. In each case A_1 is the image of A.

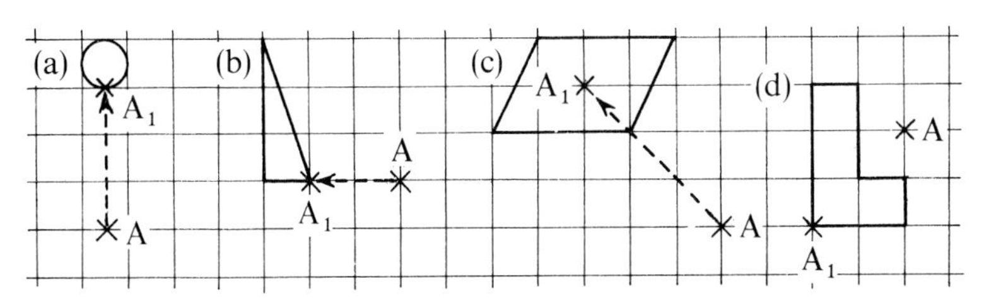

Fig. 5

6. State the distance and direction of each translation in Fig. 6.

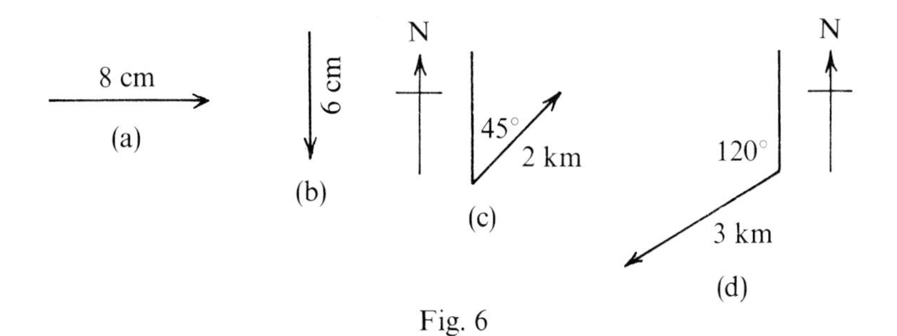

Fig. 6

7. Using suitable scales, draw arrows to represent the following translations:

(i) 7 cm upwards (ii) 5 cm to the left
(iii) 1 km on a bearing of 135°.

8. Which of the translations in Fig. 7 are the same?

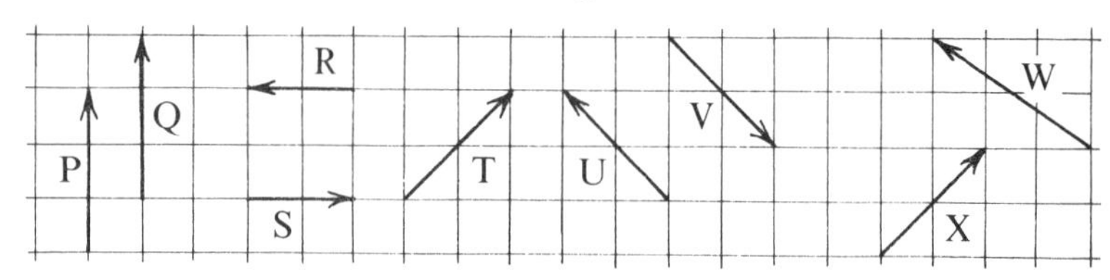

Fig. 7

9. Which of these statements about translations are true?

(i) All points on the object move the same distance in the same direction.
(ii) At least one point on the object does not move.
(iii) The image of a straight line is a parallel straight line of the same length.
(iv) The object and image have the same size and shape.

10. Draw arrows to represent two translations which move an object:

(i) In the same direction but different distances.
(ii) Equal distances in opposite directions.
(iii) Unequal distances in different directions.
(iv) Equal distances in the same direction.

11. (i) Which of the patterns in Fig. 8 can be made by translations of the unit?

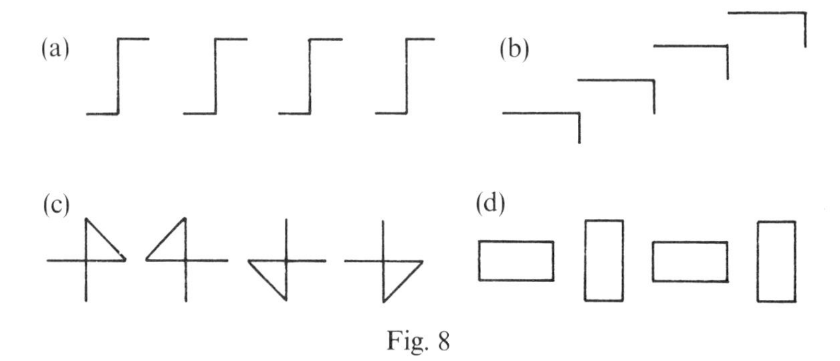

Fig. 8

(ii) Look for patterns on wallpaper, curtains and floor coverings which use the idea of translation.

12. Fig. 9 shows four identical triangles H, H_1, K, K_1. H_1 is the image of H under translation T. K_1 is the image of K under a translation in the same direction as T but for twice the distance. We say this is translation 2T.

On squared paper draw a triangle like H and show its images under the translations:

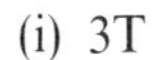 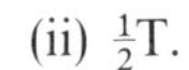

(i) 3T (ii) $\frac{1}{2}$T.

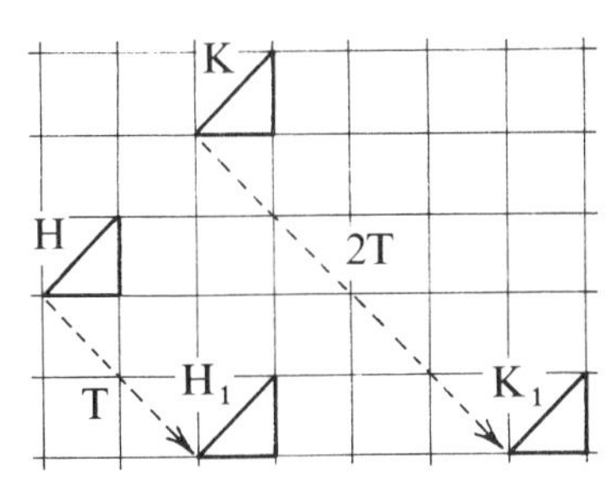

Fig. 9

SUMMARY

1. A translation is a movement in a straight line with no turning.
2. A translation has magnitude and direction.
3. Under a translation an object and its image have the same size and shape.

REFLECTIONS

In Book 2 we studied bilateral (or mirror) symmetry. When an object is placed in front of a mirror, the object and its image have bilateral symmetry, The mirror is the axis of symmetry.

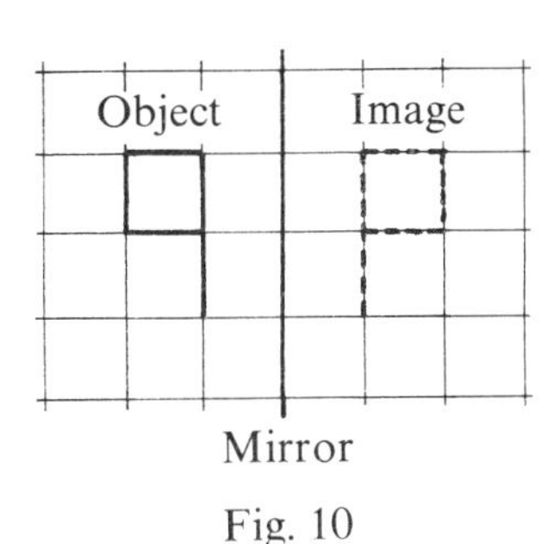

Fig. 10

Exercise 2

1. For each part of Fig. 11 say whether or not the shape I is the reflection of the shape O in the mirror M.

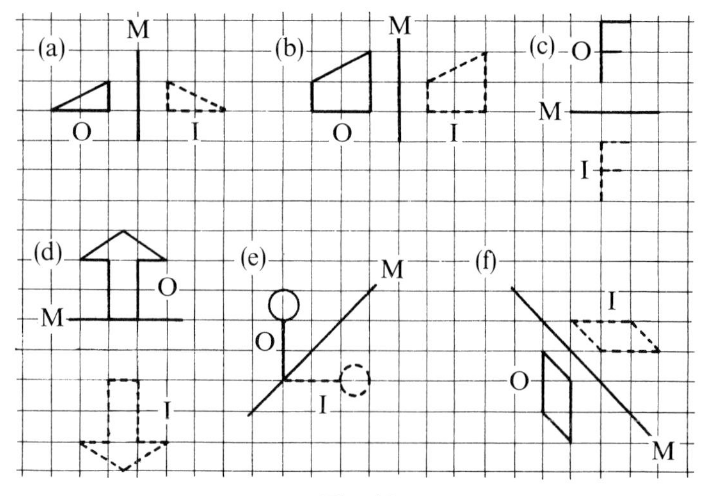

Fig. 11

2. Copy the shapes in Fig. 12 onto squared paper and draw the reflection of each in the given mirror M.

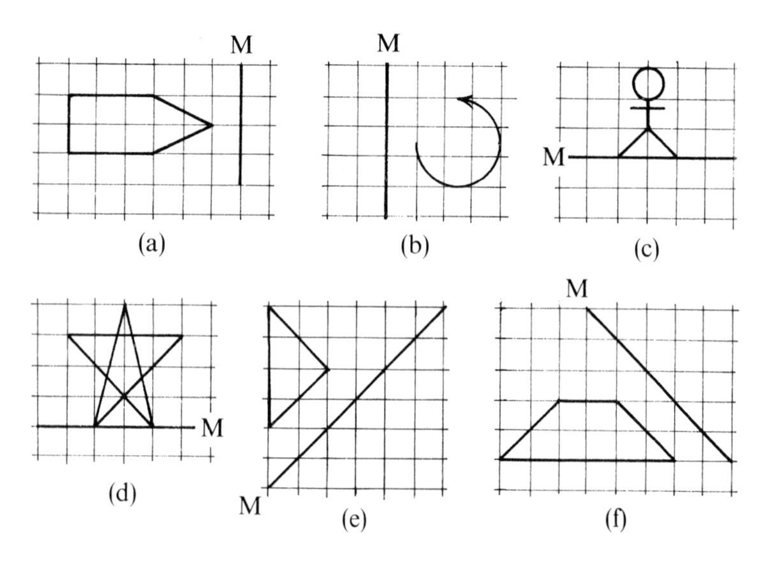

Fig. 12

3. Fig. 13 shows objects and their images formed by reflection in mirrors. Copy each diagram onto squared paper and add the mirrors.

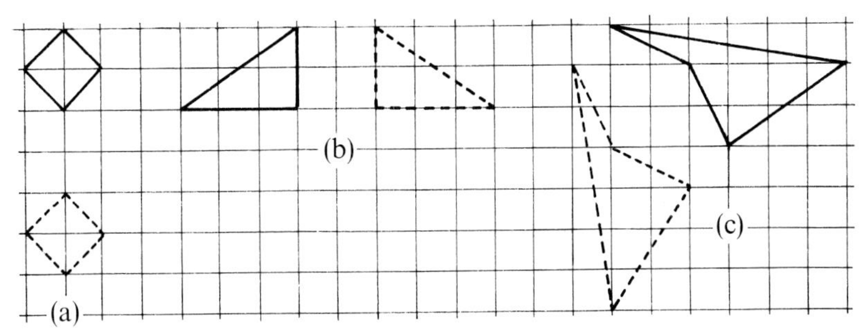

Fig. 13

4. Copy Fig. 14 on squared paper. Mark point P(3, 2). Mark as P_1 the image of P under reflection in the x axis and state its coordinates. Also mark as P_2, P_3 and P_4 the images of P under reflection in the y axis, the line $x = 1$ and the line $y = 2$. State the coordinates of these points.

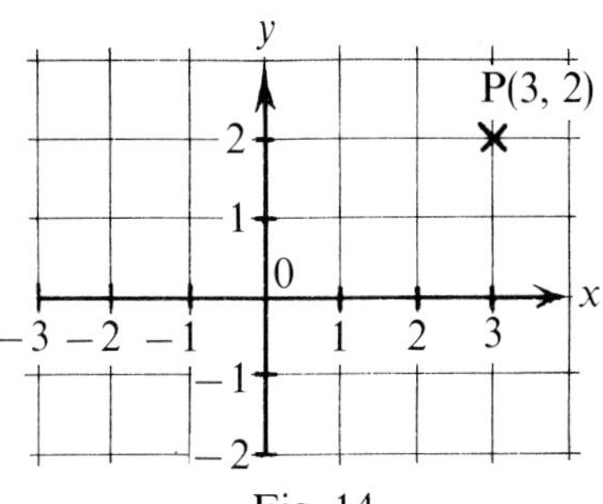

Fig. 14

5. Copy Fig. 14 again and draw the line $x = 3$ through P. Now draw the images of the line $x = 3$ under reflection in each of the mirrors used in Question **4**. State the equations of the images.

6. On squared paper draw axes as in Fig. 15 and the line AB. Using the x axis as mirror, mark A_1 and B_1 as the images of A and B. Find where the line AB and its image intersect. (You will have to extend both lines.)

Is this point on the axis of reflection?

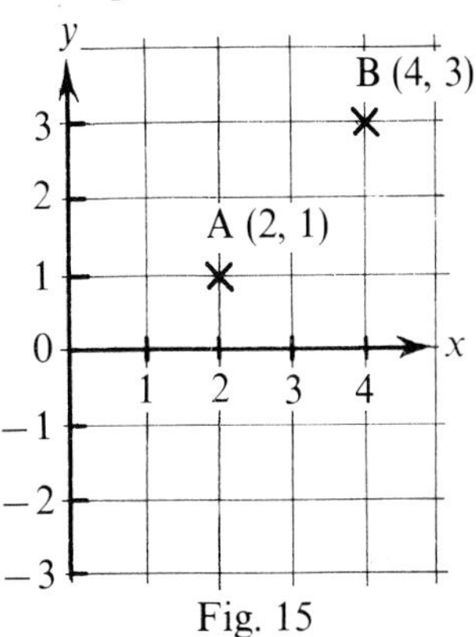

Fig. 15

7. Draw axes on squared paper and mark the x axis from -8 to $+8$ and the y axis from 0 to 10. Mark the points P(2, 5), Q(4, 1) and R(8, 2). Draw $\triangle$PQR. Draw the reflection of $\triangle$PQR in the y axis and label it $P_1Q_1R_1$. On your diagram show that:
(a) QR and its image Q_1R_1 meet at (0, 0) on the mirror.
(b) RP and its image R_1P_1 meet at (0, 6) on the mirror.
Where do PQ and its image P_1Q_1 meet?

8. Look at your drawing of Question **7**.
(i) Is the axis of reflection the mediator (perpendicular bisector) of the line joining point P to its image P_1?
(ii) Is PQ equal in length to P_1Q_1?

9. Describe in your own words how to draw the reflection, in a mirror, of:
(i) a point P (ii) a line PQ.

10. Copy Fig. 16 on squared paper. Draw the image OP_1 of line OP in the x axis. By measurement show that OP_1 and OP are inclined to the mirror Ox at the same angle of 45°. This suggests that the mirror bisects angle POP_1. Draw the image OQ_1 of line OQ in the x axis.

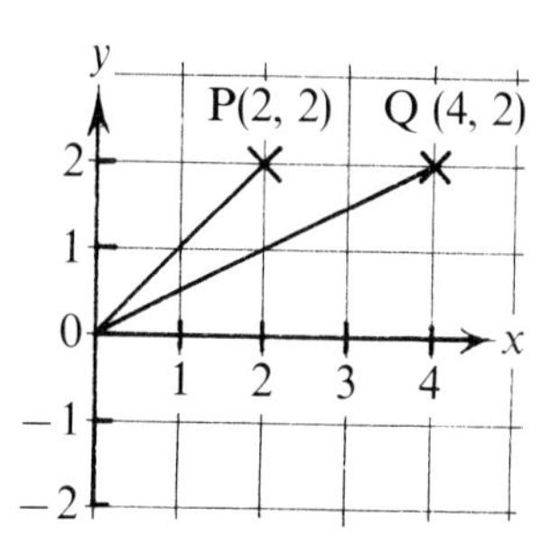

Fig. 16

Show that the axis of reflection bisects angle QOQ_1.

11. Copy Fig. 17. The line $y = x$ is to be used as the mirror or axis of reflection. Mark the images in this mirror of the points P(2, 0), Q(4, 0), R(4, 2) and S(3, 2).
Does this suggest that the image of (a, b) is (b, a)?

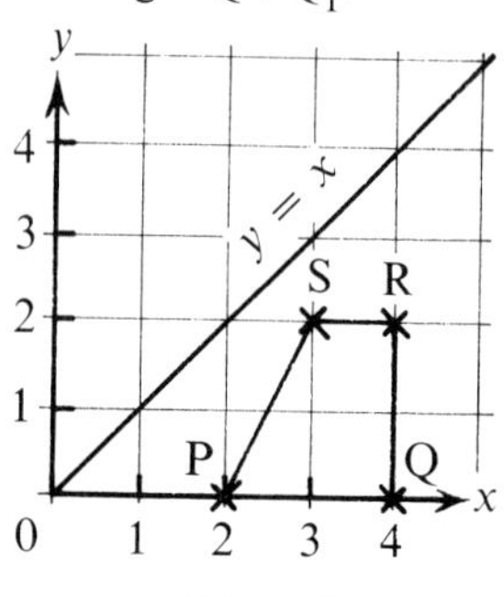

Fig. 17

12. Which statements are true about reflection in a mirror?
(i) An object and its image have the same shape and size.

(ii) An object and its image form a figure with bilateral symmetry.
(iii) The image of a straight line is a parallel straight line.
(iv) Reflection leaves distances and angles unchanged.
(v) If P_1 is the image of P, the mirror is the mediator of PP_1.
(vi) The mirror bisects the angle formed by a line and its image.

SUMMARY

1. Reflection in a mirror (axis of reflection) changes position but keeps (or preserves) size and shape.
2. The line joining a point P to its reflection P_1 is bisected by the mirror and is at right angles to the mirror.
3. An object and its image in a mirror form a figure which has bilateral symmetry.

ROTATIONS

Fig. 18 shows the turning actions for a tap, a door handle, a gate and a key. In each case there is rotation through a certain angle about a fixed axis. The axis of rotation for each object is marked by a broken line.

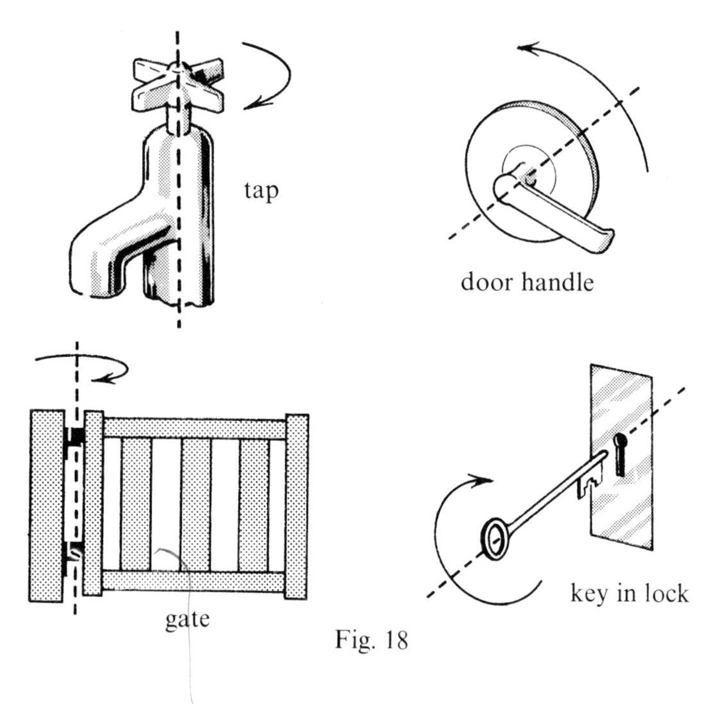

Fig. 18

Each axis of rotation is perpendicular to the plane in which the turning takes place.

Exercise 3

1. (i) Draw a line OP. With centre O and radius OP make an arc, PP_1, of a circle. Join OP_1 (Fig. 19). Measure the angle POP_1. Suppose that it is 40°. The line OP has been rotated through an angle of 40°. Line OP_1 is the image of line OP.

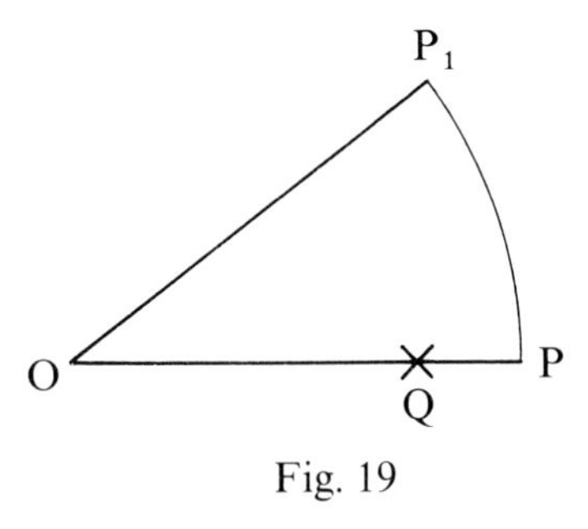

Fig. 19

The path from P to P_1 is an arc of a circle and P_1 is the image of P.

(ii) Let Q be any point on OP. With centre O and radius OQ construct its path to obtain its image Q_1.

2. Place your ruler on the desk. Hold one end with a finger and slowly rotate the ruler through an angle, using the finger as a pivot. In Fig. 20, the image of the ruler is shown by broken lines. Notice that

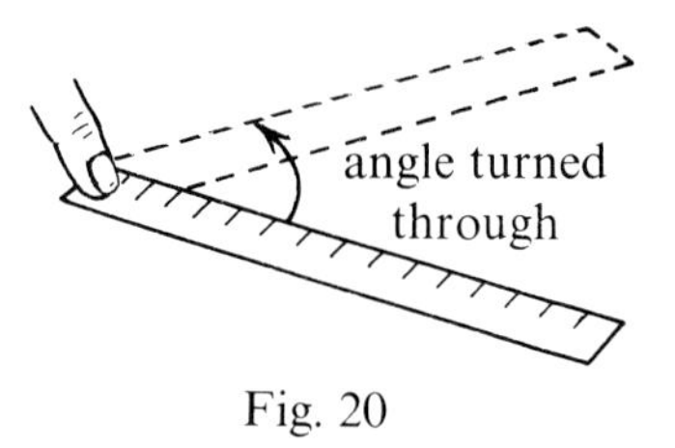

Fig. 20

(i) every part of the ruler turns through the same angle,
(ii) the point of contact of finger and ruler is the centre of rotation,
(iii) the centre of rotation does not move.

3. Repeat the work of Question **2** but use the centre of the ruler as a pivot or centre of rotation. Notice that as the ruler is turned in an anticlockwise direction, one end moves away from you and the other end moves towards you.

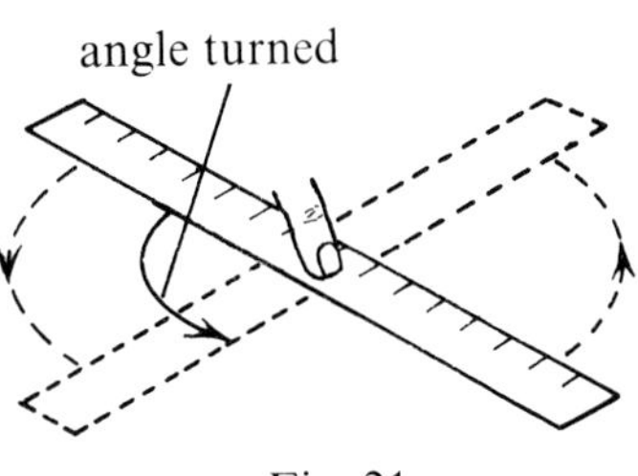

Fig. 21

4. Place a rectangular piece of card on paper and mark its outline (Fig. 22). Hold the corner A with a finger or a pencil and slowly rotate the rectangle anticlockwise through an angle of 90°.

Mark the outline of the new position $A_1B_1C_1D_1$.

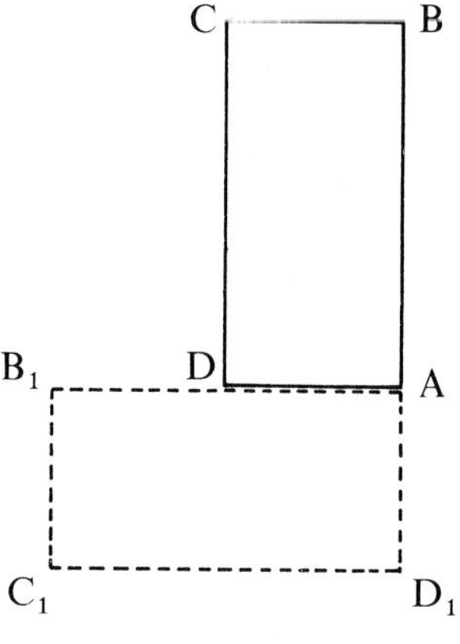

Fig. 22

Draw diagonals AC and A_1C_1. By measuring, show that angle CAC_1, the angle of rotation, is 90°.

5. Place a piece of card in the shape of a right-angled triangle on paper and mark its outline ABC (Fig. 23). Rotate it through 90° anticlockwise about A and mark its image as $A_1B_1C_1$.

Measure the angle between the hypotenuse BC and its image B_1C_1. This is 90°, the angle of rotation.

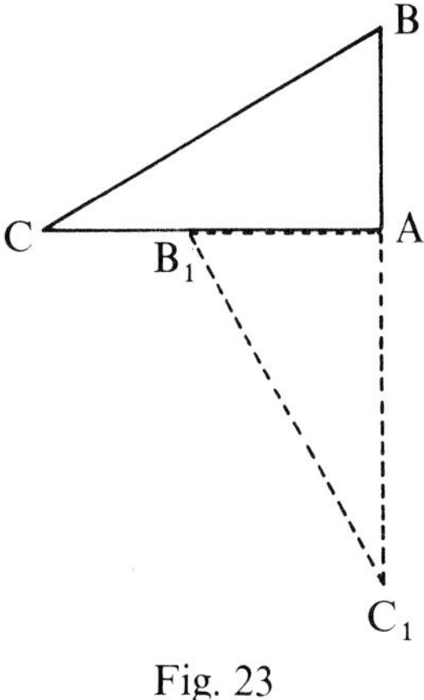

Fig. 23

On your diagram use compasses to show the circular arcs followed by B and C.

6. Place a set square on paper and draw round it. (Fig. 24.) Turn it through a convenient angle about corner P. Mark its image as $P_1Q_1R_1$. Show, by measuring, that the angles between:

(i) PQ and PQ_1
(ii) PR and PR_1
(iii) QR and Q_1R_1

are all the same. This is the angle of rotation.

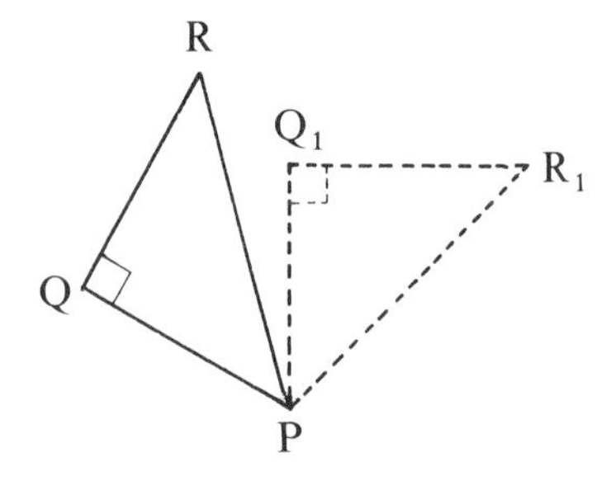

Fig. 24

7. Draw OA_1 to be the image of line OA after a rotation through 90° about O (Fig. 25). Construct the mediator (perpendicular bisector) of AA_1 and show that it passes through point O.

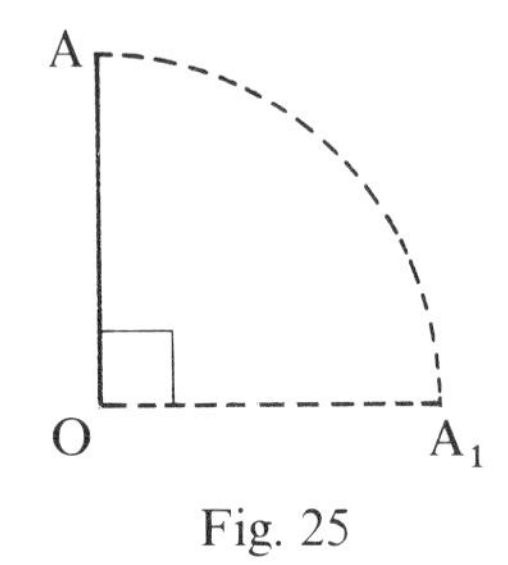

Fig. 25

8. Repeat the work of Question **7** using an angle of 60°.

9. Place a set square on paper and draw round it. (Fig. 24.) Turn it through a suitable angle using P as centre of rotation. Mark its image PQ_1R_1. Construct the mediator of QQ_1 and of RR_1. Show that they meet at P, the centre of rotation.

10. On squared paper draw lines AB and A_1B_1 as shown in Fig. 26. A_1B_1 is the image of AB under rotation. Construct the mediators of AA_1 and BB_1. Show that they meet at O, which is the centre of rotation.

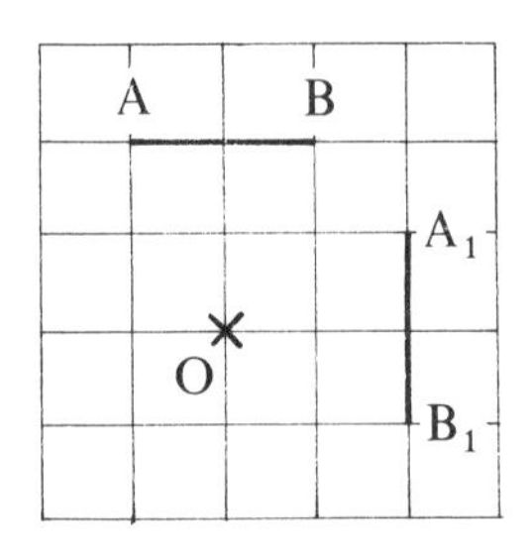

Fig. 26

By measuring angle AOA_1, show that the angle of rotation is 90°.

11. Carry out the work of Question **10** using the lines shown in Fig. 27.

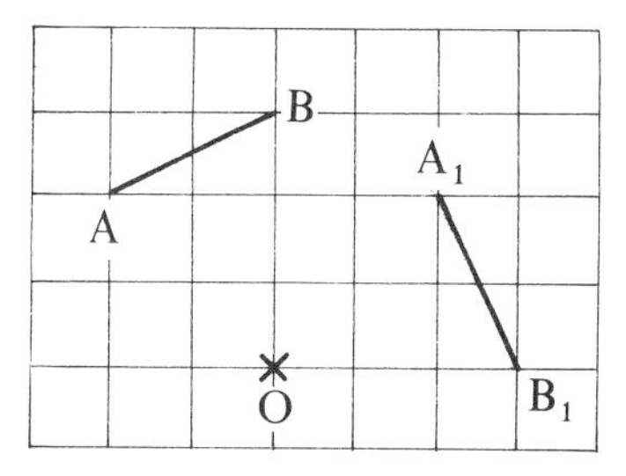

Fig. 27

12. Fig. 28 shows two identical squares ABCD and $A_1B_1C_1D_1$, each of side 5 units. Copy the diagram on squared paper.

Regard $A_1B_1C_1D_1$ as the image of ABCD under a rotation.

By constructing suitable mediators, show that O is the centre of rotation. By measuring, show that the angle of rotation is about 53°

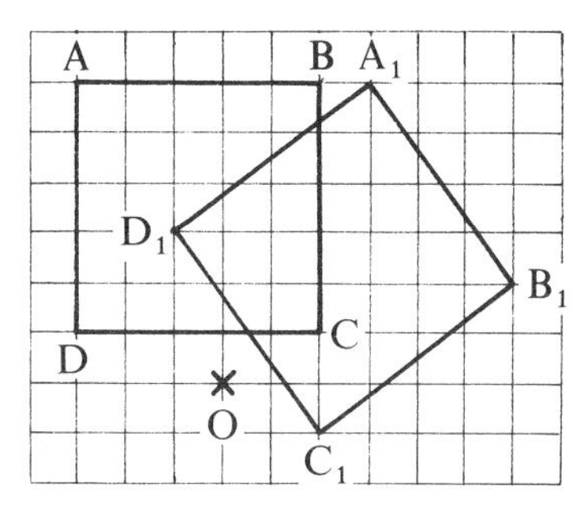

Fig. 28

SUMMARY

1. Rotation takes place in a plane about a fixed point, the centre of rotation.
2. Under a rotation all points of an object move along arcs of circles and all lines turn through the same angle.
3. The object and its image have the same size and shape.
4. If A_1 is the image of A, then the mediator of AA_1 passes through the centre of rotation.

ENLARGEMENTS

When a translation, a reflection or a rotation is applied to an object its shape and size remain the same. When the transformation of enlargement is applied to an object its size is changed but *not* its shape.

This is seen in a cinema when a small picture on film is projected as a large picture on the screen. A camera has the reverse effect. It records a large picture as a small one.

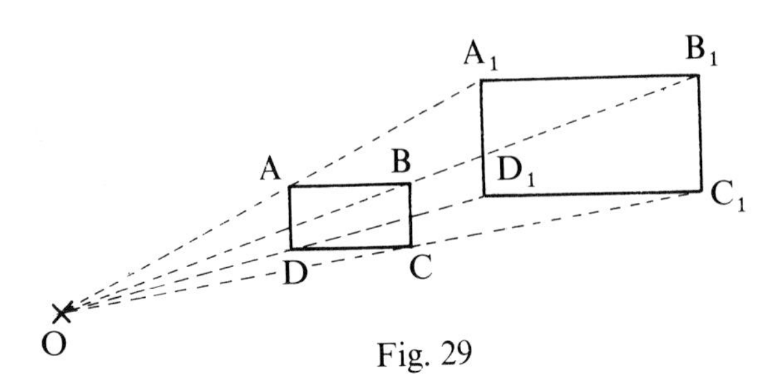

Fig. 29

Fig. 29 shows the rectangle ABCD enlarged into rectangle $A_1B_1C_1D_1$. The point O is called the *centre of enlargement*. $A_1B_1C_1D_1$ is the image of ABCD.

Copy Fig. 29. First draw rectangle ABCD 2 cm by 1 cm and choose a convenient position for point O. Join O to each corner. Make $OA_1 = 2OA$, $OB_1 = 2OB$, etc and draw $A_1B_1C_1D_1$. Measure the length and width of $A_1B_1C_1D_1$.

The dimensions of $A_1B_1C_1D_1$ are double those of ABCD.

We say that this enlargement has a *scale factor* of 2.

The two rectangles have the same shape. They are similar.

Exercise 4

1. Fig. 30 shows an enlargement of line AB using a scale factor of 3. O is the centre of enlargement. Copy the figure on squared paper. A_1B_1 is the image of AB. $A_1B_1 = 3AB$ and A_1B_1 is parallel to AB.

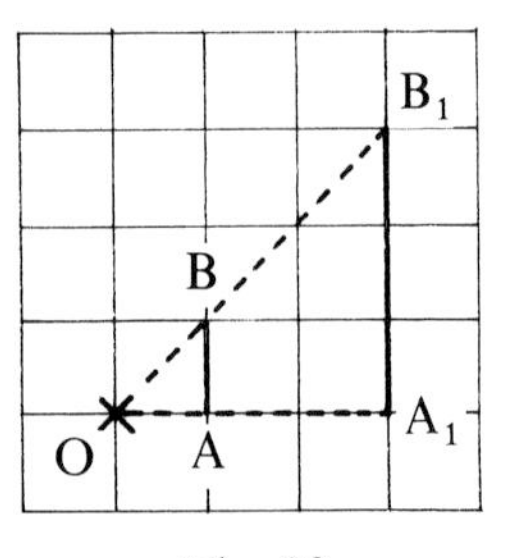

Fig. 30

2. On squared paper make three separate diagrams similar to Fig. 30 to show enlargements of line AB by scale factors of:

(i) 2 (ii) 4 (iii) 5.

In each case confirm that $A_1B_1:AB$ is the scale factor and that A_1B_1 is parallel to AB.

3. On squared paper draw triangle PQR where P is (1, 0), Q is (1, 1) and R is (2, 1). With O as centre of enlargement and a scale factor of 3, construct the image of the triangle.

4. Model aircraft are made on a scale of 1:72. What does this mean? If a model is 30 cm long, how long is the actual aircraft?

5. Model cars are made on a scale of 1:36. If an actual car is 360 cm long and 162 cm wide, what are the length and width of the model?

6. Draw a square of convenient size. Draw its diagonals and mark their point of intersection as C. Construct the image of the square using an enlargement of scale factor 2 with C as the centre of enlargement. On a separate diagram apply the same enlargement to a circle using its centre as the centre of enlargement.

7. Copy Fig. 31 on squared paper. O is the centre of enlargement and A_1B_1 is the image of AB. In this case $A_1B_1 = \frac{1}{2}AB$. The scale factor is $\frac{1}{2}$ and the image is smaller than the object.

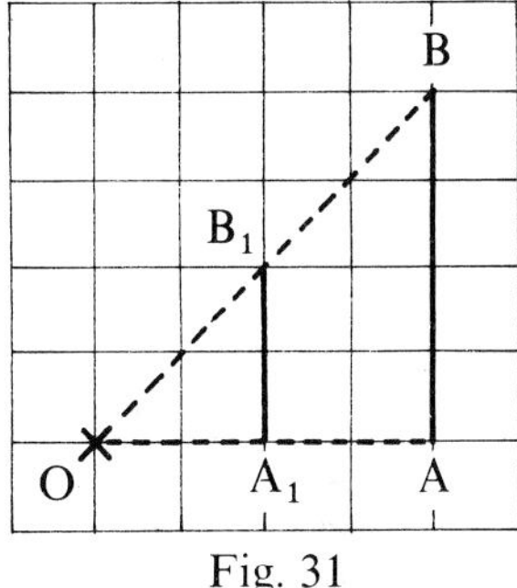

Fig. 31

8. Copy each part of Fig. 32 and construct the image of line AB using the given point O as the centre of enlargement and the given scale factor.

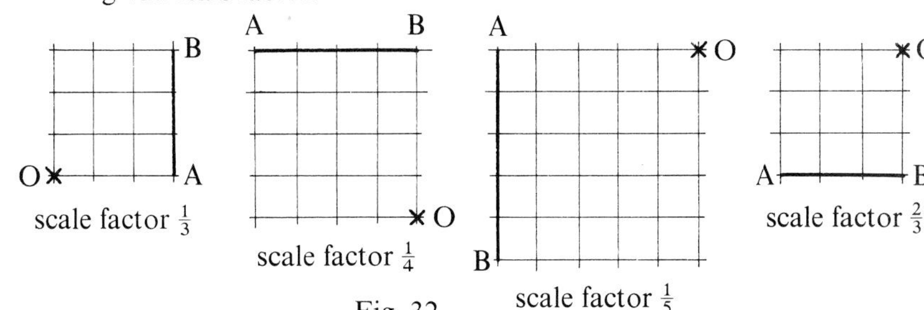

Fig. 32

9. Copy the parts of Fig. 33 and for each part construct the image of the shape using the given point O as centre of enlargement and the given scale factor.

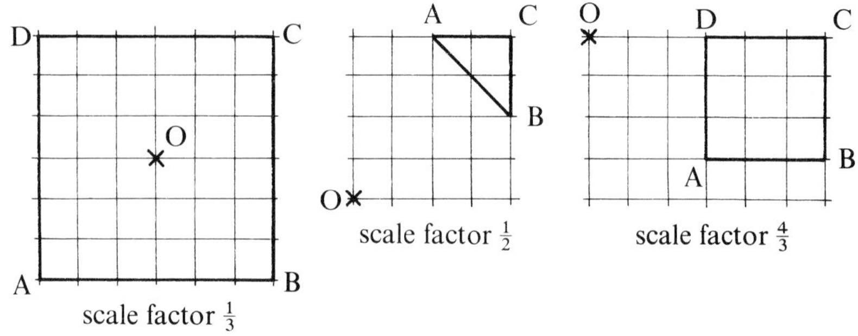

Fig. 33

SUMMARY

1. An enlargement is a transformation which uses a fixed point, O, as the centre of enlargement and a certain scale factor.

2. If A_1 is the image of point A, then OAA_1 is a straight line and the scale factor is $OA_1 : OA$.

3. The line AB has an image line A_1B_1 which is parallel to it.

4. An enlargement applied to an object changes its size but not its shape.

2 · INDICES AND STANDARD FORM

Exercise 5

1. $2^4 = 2 \times 2 \times 2 \times 2 = 16$.
Work out the value of: $2^5, 2^6, 2^7, 2^8, 2^9$ and 2^{10}.
For 2^{10} you should get 1024.

2. Work out the value of: $3^2, 3^3, 3^4, 3^5, 3^6, 3^7$ and 3^8.
3^8 is 6561.

3. Work out the value of: $4^2, 5^3, 6^3, 7^2, 10^3, 10^4$ and 10^6.

4. $625 = 5^4$. Express in this way: 9, 25, 125, 27, 81 and 100 000.

5. $5^3 \times 5^4 = (5 \times 5 \times 5) \times (5 \times 5 \times 5 \times 5)$
$= 5 \times 5 \times 5 \times 5 \times 5 \times 5 \times 5 = 5^7$
Apply this method to simplify:
$2^3 \times 2^2$, $3^2 \times 3^3$, $5^4 \times 5^2$, 7×7^3, $6^4 \times 6$, $4^2 \times 4^6$.

6. $a^3 \times a^2 = (a \times a \times a) \times (a \times a) = a \times a \times a \times a \times a = a^5$.
Apply this method to simplify:
$b^2 \times b^3$, $c^2 \times c^4$, $d^3 \times d$, $e^5 \times e^2$, $f^3 \times f^3$

7. Study the working in Question **6**. You should now be able to write down the answer to $g^7 \times g^3$ as g^{10}. Write down the answers to:
$h^9 \times h^6$, $k^{20} \times k^9$, $m^{10} \times m^{32}$, $n^7 \times n^{42}$, $p^{15} \times p^7$.

8. Simplify: $2^8 \times 2^4$, $3^5 \times 3^{10}$, $4^{12} \times 4^4$, $5^{10} \times 5^7$, $6^8 \times 6^{11}$.

9. $7^5 \div 7^3 = \dfrac{7 \times 7 \times 7 \times 7 \times 7}{7 \times 7 \times 7} = 7 \times 7 = 7^2$.

Simplify in this way:
$3^5 \div 3^2$, $5^6 \div 5^2$, $6^8 \div 6^5$, $8^4 \div 8$, $9^7 \div 9^6$.

10. Simplify: $a^6 \div a^4$, $b^7 \div b^3$, $c^7 \div c$, $d^5 \div d^4$, $e^8 \div e^2$.

11. Study the working in Questions **9** and **10**. Now write down the answers to

$2^8 \div 2^2$, $3^{10} \div 3^3$, $4^{15} \div 4^{12}$, $5^{20} \div 5^4$, $10^8 \div 10^7$.

12. Simplify: $f^{10} \div f^3$, $g^9 \div g^6$, $h^{12} \div h^{11}$, $k^{50} \div k^{10}$, $n^{14} \div n^4$.

RULES FOR INDICES

The above exercise illustrates the following two rules:

1. $a^m \times a^n = a^{m+n}$

2. $a^m \div a^n = a^{m-n}$ provided $m > n$.

EXAMPLES: $c^5 \times c^8 = c^{5+8} = c^{13}$

$d^{10} \div d^3 = d^{10-3} = d^7$

$7^5 \div 7 = 7^5 \div 7^1 = 7^{5-1} = 7^4$

$p^4 \div p = p^4 \div p^1 = p^{4-1} = p^3$

We cannot apply this method if the base numbers or letters are different as in $a^5 \times b^3$. All we can do here is to leave out the multiplication sign and write a^5b^3.

But $a^5 \times b^3 \times b^4 = a^5 \times b^7 = a^5b^7$

Also $c^2d \times c^3d^5 = c^2 \times d \times c^3 \times d^5$

$= c^2 \times c^3 \times d \times d^5$

$= c^5 \times d^6 = c^5d^6$.

We cannot leave out the multiplication sign in $5^7 \times 2^3$.

Exercise 6

1. Simplify: $3^4 \times 3^6$, $5^7 \times 5^2$, $6^5 \times 6^7$, 7×7^3, $10^3 \times 10^9$

2. Simplify: $a^3 \times a^2$, $b^4 \times b^5$, $c^2 \times c$, $d^5 \times d^8$, $e^9 \times e^6$.

3. Simplify: $2^8 \div 2^5$, $3^4 \div 3$, $5^{10} \div 5^3$, $8^9 \div 8^8$, $10^{11} \div 10^6$.

4. Simplify: $f^7 \div f^2$, $g^6 \div g^3$, $h^8 \div h$, $k^7 \div k^6$, $m^{10} \div m^2$.

5. Simplify as far as possible:

$a \times a^4$, $b^2 \times c^3$, $d^6 \times d^5$, $e^{10} \times f^2$, $g^4 \times g^7$.

6. Simplify where possible:
$2^4 \times 2^5$, $3^2 \times 5^3$, $7^2 \times 7^3$, $11^4 \times 11$, $2^3 \times 5^4$.

7. Which one of the following is equal to $h^5 \times g^3$?
g^8, g^{15}, hg^8, $(hg)^{15}$, g^3h^5.

8. Which one of the following is equal to $p^8 \div r^2$?

p^6, p^4, $\frac{p^8}{r^2}$, $\left(\frac{p}{r}\right)^4$.

9. Simplify:
(i) $4^3 \times 4^2 \times 4^2$ (ii) $3^5 \times 3 \times 3^2$ (iii) $5 \times 5^2 \times 5^3$
(iv) $a^3 \times a^3 \times a^2$ (v) $b \times b^3 \times b^2$ (vi) $c \times c \times c^7$.

10. Simplify:
(i) $a^2 \times b^3 \times b^4$ (ii) $c^2 \times c^3 \times d^4$ (iii) $e^2 \times e^3 \times e^4$
(iv) $f^2 \times g^3 \times h^4$ (v) $k \times k^5 \times m$ (vi) $n \times p^3 \times p$.

11. Simplify:
(i) $a^2 \times a^3b$ (ii) $c^3d \times d^4$ (iii) $ef \times f$
(iv) $g^5 \times gh^3$ (v) $km^2 \times k^2m$ (vi) $n^3p \times np$.

12. $5^9 \div 5^7 = 5^{9-7} = 5^2 = 5 \times 5 = 25$. Use this method to work out the value of:
(i) $3^9 \div 3^7$ (ii) $2^8 \div 2^5$ (iii) $4^7 \div 4^6$ (iv) $2^{10} \div 2^8$.

13. Find the value of x for each of the following statements:
(i) $2^x = 16$ (ii) $10^x = 100$ (iii) $3^x = 243$.

14. Find the value of y for each of the following:
(i) $3^y \times 3^2 = 3^6$ (ii) $5^4 \times 5^y = 5^{12}$ (iii) $7^y \div 7^2 = 7^6$.

15. $(a^5)^3$ means $a^5 \times a^5 \times a^5 = a^{5+5+5} = a^{15}$
Simplify in this way: $(b^4)^2$, $(c^5)^2$, $(2^7)^3$ and $(7^5)^4$.

TWO MORE RULES

Consider the following:

$$(7^2)^3 = 7^2 \times 7^2 \times 7^2 = 7^{2+2+2} = 7^{3\times2} = 7^6$$
$$(q^5)^4 = q^5 \times q^5 \times q^5 \times q^5 = q^{5+5+5+5} = q^{4\times5} = q^{20}$$
$$(t^7)^2 = t^7 \times t^7 = t^{7+7} = t^{2\times7} = t^{14}$$

These are examples of:

Rule **3**: $(a^m)^p = a^{mp}$

We can write the fraction $\frac{5}{35}$ in a simpler form by dividing the numerator and denominator by 5 to obtain $\frac{1}{7}$.

Similarly $p^3 \div p^7 = \dfrac{p^3}{p^7} = \dfrac{p^3 \div p^3}{p^7 \div p^3} = \dfrac{1}{p^4}$

$$m^2 \div m^5 = \frac{m^2}{m^5} = \frac{m^2 \div m^2}{m^5 \div m^2} = \frac{1}{m^3}$$

These are examples of:

Rule **4**: If $m < n$, $a^m \div a^n = \dfrac{1}{a^{n-m}}$

Exercise 7

1. Simplify: $(a^3)^2, (b^5)^3, (c^4)^4, (d^6)^3, (e^7)^2$.

2. Simplify: $(2^4)^3, (3^2)^5, (7^3)^2, (6^5)^3, (5^5)^4$.

3. Simplify: $f^2 \div f^6, g^3 \div g^{10}, h \div h^4, k^5 \div k^5, m^2 \div m^{10}$.

4. $5^9 \div 5^{11} = \dfrac{1}{5^{11-9}} = \dfrac{1}{5^2} = \dfrac{1}{25}$

Using this method, find the value of:
$2^3 \div 2^5$, $\quad 3^4 \div 3^7$, $\quad 4^2 \div 4^3$, $\quad 2^5 \div 2^8$, $\quad 10^3 \div 10^5$.

5. Find the value of:
$2^9 \div 2^7$, $\quad 2^7 \div 2^4$, $\quad 3^8 \div 3^7$, $\quad 5^8 \div 5^5$, $\quad 10^9 \div 10^7$.

6. (i) Copy and complete: $(6^4 \times 6^3) \div 6^9 = 6^{\cdots} \div 6^9 = \dfrac{1}{6^{\cdots}} = \dfrac{1}{36}$

(ii) Find the value of:
$(2^3 \times 2^4) \div 2^{10}$, $\quad (3^2 \times 3) \div 3^5$,
$(4^3 \times 4^4) \div 4^7$, $\quad 5^6 \div (5^4 \times 5^3)$.

7. Which of the following are true?

(i) $a^3 + a^4 = a^7$ (ii) $b^{10} \div b^2 = b^5$
(iii) $(c^3)^4 = c^{12}$ (iv) $d^2 \times d^4 = d^6$
(v) $e^7 \div e^9 = e^2$ (vi) $(f^4)^3 = f^7$
(vii) $g \times g^5 = g^5$ (viii) $h^5 \div h = h^4$
(ix) $x^7 - x^5 = x^2$ (x) $y^8 \div y^2 = y^6$.

STANDARD FORM FOR LARGE NUMBERS

Scientists use very large numbers and very small numbers. For example, the distance of Alpha Centauri, the nearest star (apart from the Sun), is 42 000 000 000 000 kilometres and the mass of an electron is 0·000 000 000 000 000 000 000 000 000 91 grammes.

When we write such numbers in this way, we need long lines of noughts and it is difficult to avoid mistakes. There is a better way which uses powers of 10.

$$42\,000\,000\,000\,000 = 4.2 \times 10\,000\,000\,000\,000 = 4.2 \times 10^{13}$$

Similarly

$$2\,578\,000 = 2.578 \times 1\,000\,000 = 2.578 \times 10^6$$

When written as 4.2×10^{13} and 2.578×10^6 the numbers are said to be in *standard form*.
For standard form we use

1. a number between 1 and 10, together with,

2. a power of 10.

That is, we use $a \times 10^n$ where $1 \leqslant a < 10$ and n is an integer.

Here are three more examples:

$$90\,000 = 9 \times 10\,000 = 9 \times 10^4$$
$$560 = 5.6 \times 100 = 5.6 \times 10^2$$
$$3467 = 3.467 \times 1000 = 3.467 \times 10^3$$

Exercise 8

1. Which of the following numbers are in standard form:
(i) 3.9×10^5 (ii) 63×10^4 (iii) 5×10^8 (iv) 7×5^4?

2. Write in standard form
(i) 800 (ii) 5000 (iii) 70 000 (iv) 4 000 000.

3. Write the following without powers of 10:
(For example: $7 \times 10^3 = 7 \times 1000 = 7000$)
(i) 6×10^2 (ii) 3×10^4 (iii) 9×10^6 (iv) 2×10^5.

4. Write in standard form:

(i) 350 (ii) 7600 (iii) 132 000 (iv) 68 000
(v) 941 (vi) 3625 (vii) 75 320 (viii) 85 200 000.

5. Write without powers of 10:

(i) 4.71×10^2 (ii) 6.837×10^2 (iii) 5.62×10^3
(iv) 7.6×10^4 (v) 9.54×10^6 (vi) 4.09×10^4.

6. A billion is a million million and a trillion is a million billion. Write in standard form:

(i) five million (ii) three billion
(iii) seven trillion (iv) two hundred trillion.

7. Write in standard form:

(i) The mass of the Earth, 5980 trillion tonnes.
(ii) The distance of Mars from the Sun, 228 million kilometres.
(iii) The height of Mount Everest, 8900 metres.
(iv) The speed of light, 300 000 kilometres per second.

8. Calculate:

(i) the number of centimetres in a kilometre,
(ii) the number of cubic centimetres in a cubic metre, giving each answer as a power of 10.

9. Calculate the time taken to count up to a million at the rate of one number each second. Given the answer to 2 significant figures in the most reasonable unit of time.

10. Calculate the number of seconds in a year. Give the answer in standard form, correct to 2 sig. fig. How many seconds have you lived, correct to 2 sig. fig.?

11. A 'light year' is the distance travelled by light in one year. Calculate, correct to 2 sig. fig., the distance in kilometres equal to one light year (see Question 7, part (iv)).

The distance of the brightest star, Sirius, is 8.6 light years. Express this in kilometres to 2 sig. fig.

12. $537 \times 10^3 = 5.37 \times 100 \times 10^3 = 5.37 \times 10^2 \times 10^3$
$= 5.37 \times 10^5$

Use this method to put the following numbers into standard form:

(i) 63×10^4 (ii) 72.9×10^3 (iii) 540×10^5
(iv) 233×10^2 (v) 9600×10 (vi) 36.5×10^4

13. $(6 \times 10^3) \times (3 \times 10^4) = 6 \times 3 \times 10^3 \times 10^4 = 18 \times 10^7$
$= 1.8 \times 10 \times 10^7 = 1.8 \times 10^8$

Do the following calculations in this way, giving each answer in standard form:

(i) $(4 \times 10^2) \times (7 \times 10)$ (ii) $(9 \times 10^4) \times (4 \times 10^3)$
(iii) $(5 \times 10^2) \times (6 \times 10^3)$ (iv) $(4.8 \times 10^4) \times (1.5 \times 10^2)$
(v) $600 \times 840\,000$ (vi) $750 \times 24\,000$.

14. $0.046 \times 10^5 = 0.046 \times 100 \times 10^3 = 4.6 \times 10^3$

Use this method to put the following numbers into standard form:

(i) 0.36×10^5 (ii) 0.8×10^7 (iii) 0.627×10^4
(iv) 0.02×10^6 (v) 0.044×10^8 (vi) 0.73×10^6.

15. $(8.4 \times 10^7) \div (3 \times 10^4) = \dfrac{8.4 \times 10^7}{3 \times 10^4} = \dfrac{8.4}{3} \times \dfrac{10^7}{10^4} = 2.8 \times 10^3$

Do the following divisions in this way, giving you answers in standard form:

(i) $(7.2 \times 10^5) \div (4 \times 10^2)$ (ii) $(9.6 \times 10^{12}) \div (3 \times 10^5)$
(iii) $(8.5 \times 10^{10}) \div (5 \times 10^6)$ (iv) $(3.5 \times 10^8) \div (7 \times 10^2)$
(v) $(5.4 \times 10^{12}) \div (6 \times 10^4)$ (vi) $(1.08 \times 10^5) \div (9 \times 10^2)$

NEGATIVE INDICES

In standard form the mass of an electron is written 9.1×10^{-28} grammes. Here we have a *negative* index. We now consider what this means. Study this table:

Number	100 000	10 000	1000	100	...	...
Index Form	10^5	10^4	10^3	10^2	...	...

What are the next six numbers in the sequence

100 000, 10 000, 1000, 100, ...?

What are the next six in the sequence 5, 4, 3, 2, ...?

The table can be continued as follows:

Number	10	1	$\frac{1}{10}$	$\frac{1}{100}$	$\frac{1}{1000}$	$\frac{1}{10\,000}$
Index Form	10^1	10^0	10^{-1}	10^{-2}	10^{-3}	10^{-4}

This suggests that we should take 10^0 to mean 1, 10^{-1} to mean $\frac{1}{10}$, 10^{-2} to mean $\frac{1}{100}$ and so on.

Notice that $\dfrac{1}{1000} = \dfrac{1}{10^3}$ so that 10^{-3} means $\dfrac{1}{10^3}$.

Similarly 10^{-5} means $\frac{1}{10^5}$, that is $\frac{1}{100\,000}$ or 0.000 01.

Likewise $2^{-3} = \frac{1}{2^3} = \frac{1}{8}$ and $a^{-7} = \frac{1}{a^7}$.

The four rules for working with positive indices also apply to negative indices. Here are some examples:

$$2^{-3} \times 2^{-4} = 2^{-3-4} = 2^{-7} \text{ since } \frac{1}{8} \times \frac{1}{16} = \frac{1}{128}$$

$$a^5 \times a^{-2} = a^{5-2} = a^3 \text{ since } a^5 \times \frac{1}{a^2} = a^3$$

$$b^3 \div b^7 = b^{3-7} = b^{-4} \text{ since } \frac{b^3}{b^7} = \frac{1}{b^4}$$

Exercise 9

1. Write as 10^n:
   100, $\frac{1}{100}$, 10 000, $\frac{1}{10\,000}$, 0.01, 1.
2. Write the following without indices. Give both the fraction and decimal form where possible: 10^3, 10^{-3}, 10^0, 10^{-1}, 10^{-4}.
3. $4^{-2} = \frac{1}{4^2} = \frac{1}{16}$. Give as fractions:
   3^{-2}, 2^{-3}, 5^{-2}, 6^{-1}, 2^{-4}.
4. $\frac{1}{81} = \frac{1}{3^4} = 3^{-4}$. Copy and complete:
   $\frac{1}{6^3} = 6^{\cdots}$, $\frac{1}{8^4} = 8^{\cdots}$, $\frac{1}{64} = \frac{1}{4^{\cdots}} = 4^{\cdots}$, $\frac{1}{125} = \frac{1}{5^{\cdots}} = 5^{\cdots}$.
5. Copy and complete this table:

Number	8	4	2	1	$\frac{1}{2}$	$\frac{1}{4}$	$\frac{1}{8}$	$\frac{1}{16}$	$\frac{1}{32}$	$\frac{1}{64}$
Power of 2	2^3	2^2	...	...	...	...	2^{-3}	...	...	...

6. Give the value of: 3^3, 3^{-3}, 3^0, 3^{-4}, 3^{-1}.
7. Express as powers of 3: 9, $\frac{1}{9}$, $\frac{1}{3}$, 1, 81.

8. Write in the form a^n: $\frac{1}{a^3}$, $\frac{1}{a^5}$, $\frac{1}{a}$, 1, $\frac{1}{a^7}$.

9. Give the answers to the following in the form 10^n:

100×1000, $10 \times \frac{1}{1000}$, $1000 \div 10$, $10 \div 100$,
$10^3 \times 10^4$, $10^{-4} \times 10^2$, $10^{-2} \times 10^5$, $10^{-3} \times 10^{-2}$
$10^5 \div 10^2$, $10^2 \div 10^5$, $10^6 \div 10^{-2}$, $10^{-1} \div 10^{-4}$

10. Simplify:

$a^{-4} \times a^{-3}$, $b^{-2} \times b^{-3}$, $c^{-4} \times c^{10}$, $d^{-8} \times d^3$
$e^6 \div e^2$, $f^2 \div f^6$, $g^{-2} \div g^{-5}$, $h^{-5} \div h^{-2}$.

STANDARD FORM FOR SMALL NUMBERS

We can now write small numbers in standard form.

$$0.06 = 6 \times 0.01 = 6 \times \frac{1}{100} = 6 \times \frac{1}{10^2} = 6 \times 10^{-2}$$

$$0.004\,72 = 4.72 \div 1000 = 4.72 \times \frac{1}{1000} = 4.72 \times \frac{1}{10^3} = 4.72 \times 10^{-3}$$

In the reverse direction:

$$6.8 \times 10^{-4} = 6.8 \times \frac{1}{10^4} = 6.8 \times \frac{1}{10\,000} = 6.8 \div 10\,000 = 0.000\,68$$

Exercise 10

1. Copy and complete:

(i) $0.05 = 5 \times 0.\ldots = 5 \times \frac{1}{\ldots} = 5 \times \frac{1}{10^{\cdots}} = 5 \times 10^{-\cdots}$

(ii) $0.0008 = 8 \times \ldots = \ldots = \ldots = \ldots$

(iii) $0.39 = 3.9 \times \ldots = \ldots = \ldots = \ldots$

(iv) $0.064 = 6.4 \times \ldots = \ldots = \ldots = \ldots$

2. Write in standard form:

0.04, 0.0007, 0.3, 0.002, 0.000 08
0.082, 0.009 36, 0.165, 0.0073, 0.0304.

3. $0.000\,56 = 5.6 \times 10^{-4}$. How many noughts are there before the 5 in 0.000 56?

What is the index in 5.6×10^{-4}?

Answer the same questions for your results in Question **2**.

What do you notice? Now write down in standard form:

0.0009, 0.0416, 0.0082, 0.714, 0.0409.

4. Write without powers of 10:

7.3×10^{-3}, 9.5×10^{-4}, 6×10^{-2}, 4.9×10^{-6}
1.68×10^{-2}, 5.7×10^{-1}, 8.06×10^{-2}, 5×10^{-4}.

5. Write in standard form: 6 thousandths, 7 millionths, 55 millionths, $23 \div 10^4$, $605 \div 10^5$, $\frac{3}{10}$.

6. Write in standard form:

(i) The wavelength of green light in the mercury spectrum, 0.000 054 6 cm.
(ii) The charge of an electron, 0.158 trillionths of a coulomb.
(iii) 24 millionths of a metre (24 micrometres).
(iv) 0.000 000 006 metre (6 nanometres).

7. Do the following calculations and give your answers in standard form:

(i) $(0.08)^2$ (ii) $(0.2)^3$ (iii) 0.6×0.045
(iv) $0.0007 \div 5$ (v) $0.075 \div 2.5$ (vi) $20 \div 0.0125$.

8. Give the answers to the following in standard form, correct to 2 sig. fig.:

(i) $0.023 \div 60$ (ii) 0.326×0.7 (iii) $\dfrac{0.4 \times 0.08}{7}$.

3 · GRAPHS OF RELATIONS

Two towns are 60 km apart on a main road. A cyclist takes 4 h to cycle from one town to the other. His average speed is $\frac{60}{4} = 15$ km/h.

What is the average speed of:

(i) a bus which takes 2 h,
(ii) a lorry which takes $1\frac{1}{2}$ h?

If the time taken is x h, the average speed, y km/h, is given by $y = \dfrac{60}{x}$.

Copy and complete the following table of values of x and y

x	1	$1\frac{1}{2}$	2	$2\frac{1}{2}$	3	4	5	6
y	60	40		24		15		

Fig. 1 shows the graph which is obtained by plotting the points (1, 60), $(1\frac{1}{2}, 40)$, etc, and drawing a curve through them.

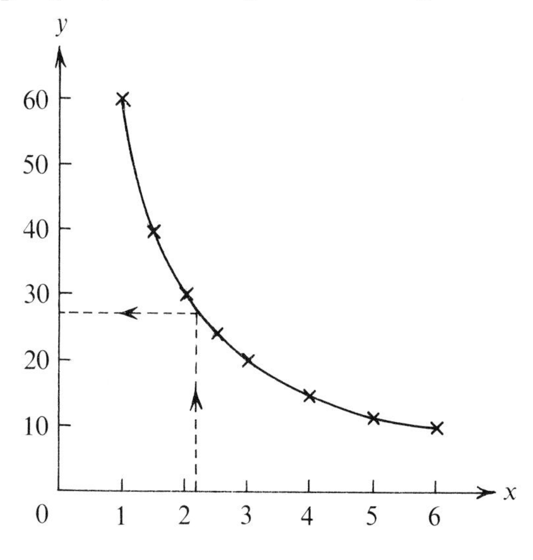

Fig. 1

Draw your own graph using 2 cm to represent 1 unit on the x-axis and 2 cm to represent 10 units on the y-axis. Make sure that your curve is smooth.

From your graph find the average speed if the time taken is 2.2 h. The broken line in Fig. 1 shows you how to do this. Also find the average speed if the time taken is 4.6 h.

Use your graph to find the time taken if the average speed is:

(i) 17 km/h (ii) 34 km/h.

In the above we have two variables—time (x h) and speed (y km/h). The relation between x and y is expressed by the formula $y = \frac{60}{x}$. The graph is a picture of this relation.

Notice that:

(i) each pair of values of x and y which fit the formula are represented by a point on the graph and,

(ii) each point on the graph represents a pair of values of x and y which fit the formula.

For example (4.8, 12.5) is a point on the graph and $x = 4.8$, $y = 12.5$ fit the formula $y = \frac{60}{x}$ since $\frac{60}{4.8} = 12.5$.

In set language, the graph of $y = \frac{60}{x}$ is

$$\left\{\text{points } (x, y) \text{ such that } y = \frac{60}{x}\right\}.$$

Of course Fig. 1 shows only part of the set. For example, it does not show (10, 6) and ($\frac{1}{3}$, 180). It really shows

$$\left\{\text{points } (x, y) \text{ such that } y = \frac{60}{x} \text{ where } 1 \leqslant x \leqslant 6\right\}.$$

What happens to the graph if it is continued to the right? Consider $x = 10$ for a hiker and $x = 100$ for a tortoise.

What happens to the graph if it is continued upwards? Consider $x = \frac{1}{2}$ for a fast car and $x = \frac{1}{4}$ for a helicopter.

Exercise 11

The questions have been planned to fit size A5 (148 mm × 210 mm) graph paper having squares of side 2 mm.

1. Copy and complete this table for the graph of $y = x^2$

x	-3	$-2\frac{1}{2}$	-2	$-1\frac{1}{2}$	-1	$-\frac{1}{2}$	0	$\frac{1}{2}$	...	3
$y = x^2$	9	$6\frac{1}{4}$	...	$2\frac{1}{4}$	...	...	...	...	...	

On graph paper draw axes as shown in Fig. 2. Use 2 cm to 1 unit on each axis. Plot the points $(-3, 9)$, $(-2\frac{1}{4}, 6\frac{1}{4})$, etc, corresponding to the values in your table.

Draw a smooth curve through the points.

What is the axis of symmetry of the curve?

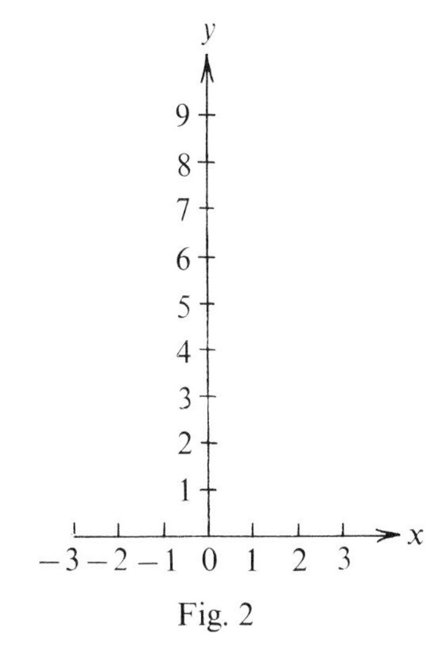

Fig. 2

The point $(2.2, b)$ is on the curve. Find an approximate value for b from your graph. It is an approximation for 2.2^2.

The point $(c, 2.6)$ is on the curve. From your graph find two possible values for c. They are approximations for the two square roots of 2.6.

Use your graph to find approximations for 1.8^2, 2.7^2 and $\sqrt{2.4}$

2. In this question the graph of $y = x^3$ is drawn. Form a table as in Question **1** to show the values of x and x^3 for $x = -3, -2.5, -2, \ldots + 3$. Remember that $(-2)^3 = -8$. Use $1.5^3 \simeq 3.4$ and $2.5^3 \simeq 15.6$.

Take axes as shown in Fig. 3, using 2 cm to 1 unit on the x axis and 2 cm to 10 units on the y axis.

Plot the points $(-3, -27)$, etc, corresponding to the values in your table and then join them with a smooth curve.

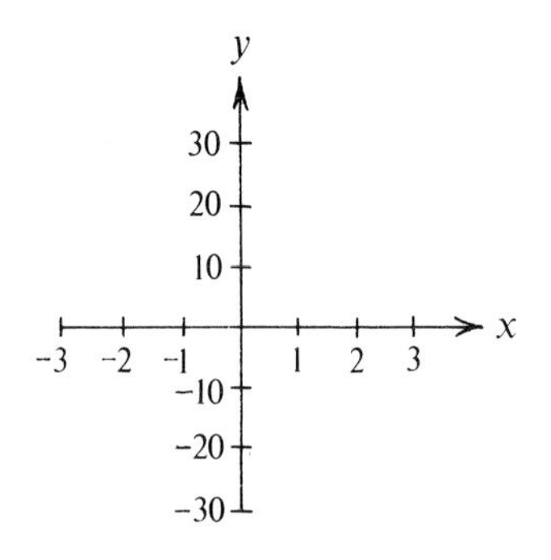

Fig. 3

Does the curve have line symmetry or point symmetry? From your graph state approximate values for: 1.8^3, 2.2^3, $\sqrt[3]{20}$ and $\sqrt[3]{(-14)}$.

3. Copy and complete this table for the graph of $y = \dfrac{120}{x}$.

x	-30	-20	-15	-12	-10	-8	-4
$y = \dfrac{120}{x}$	-4		-8				-30
x	$+4$	$+8$	$+10$	$+12$	$+15$	$+20$	$+30$
$y = \dfrac{120}{x}$	$+30$				$+8$		$+4$

Draw axes to cut in the centre of your paper as in Fig. 3. Using 2 cm to 10 units, number both axes from -30 to $+30$. Plot the points $(-30, -4)$ etc, and join them with a smooth curve.

Comment on the symmetry of the curve.

Compare the shape of this graph with Fig. 1, on page 27.

Make a simple sketch of the graph of $y = \dfrac{60}{x}$ for values of x from -1 to -6.

4. Draw the graph of $y = x^2$ as in Question **1**.

Draw also the graph of $y = \frac{1}{2}x + 3$. Remember that this should be a straight line.

State the coordinates of the points at which the graphs intersect.

5. Draw the graph of $y = x^3$ as in Question **2.**
Draw also the graphs of $y = 4x$ and $y = 10 - 5x$.
State the points at which each straight line cuts the curve.

6. State the values of $x^2 - 20$ if $x = 5, 4, -4$ and -5.
Copy and complete this table for the graph of $y = x^2 - 20$.

x	-6	-5	-4	-3	...	$+6$
x^2	36			9		
$y = x^2 - 20$	16			-11		

Number the x axis from -6 to $+6$ using 1 cm to each unit and the y axis from -20 to $+60$ using 2 cm to 10 units.
Plot the points and draw the graph.
Using the same axes draw the graphs of $y = x^2$ and $y = x^2 + 20$.
Comment on the shape of the three graphs. How can the first graph be transformed into the second and into the third?

7. State the values of $-x^2$ for $x = 5, 3, 0, -3, -5$.
Draw the graphs of $y = x^2$ and $y = -x^2$ for values of x from -6 to $+6$. Use 1 cm for 1 unit on the x-axis and 2 cm for 10 units on the y-axis which should be numbered from -40 to $+40$.
How can one curve be transformed into the other?

4 · ALGEBRAIC FRACTIONS

$\frac{2}{7}$ can be expressed as a fraction with 21 as its denominator thus:

$$\frac{2}{7} = \frac{3 \times 2}{3 \times 7} = \frac{6}{21}$$

Similarly $\frac{c}{d}$ can be expressed as a fraction with ad as its denominator thus:

$$\frac{c}{d} = \frac{a \times c}{a \times d} = \frac{ac}{ad}$$

Likewise, $\frac{v}{w}$ can be expressed as a fraction with $5w^2$ as its denominator thus:

$$\frac{v}{w} = \frac{5 \times w \times v}{5 \times w \times w} = \frac{5wv}{5w^2}$$

In the reverse direction, a fraction can sometimes be expressed with smaller numbers, that is, reduced to lower terms.

Thus $$\frac{12}{15} = \frac{3 \times 4}{3 \times 5} = \frac{4}{5}$$

Similarly $$\frac{pn}{pt} = \frac{p \times n}{p \times t} = \frac{n}{t}$$

The expression $3(k + m)$ has two factors. One factor is 3 and the other is $(k + m)$.

Hence $$\frac{3(k + m)}{7(k + m)} = \frac{3 \times (k + m)}{7 \times (k + m)} = \frac{3}{7}$$

Exercise 12

Copy and complete:

1. $\frac{1}{3} = \frac{}{12}$ **2.** $\frac{1}{a} = \frac{}{ab}$ **3.** $\frac{1}{c} = \frac{}{c^2}$ **4.** $\frac{1}{d} = \frac{}{5d}$

5. $\frac{2}{5} = \frac{}{15}$ **6.** $\frac{3}{f} = \frac{}{2f}$ **7.** $\frac{4}{g} = \frac{}{gh}$ **8.** $\frac{5}{k} = \frac{}{k^2}$

9. $\frac{m}{p} = \frac{}{pq}$ **10.** $\frac{u}{x} = \frac{}{3x}$ **11.** $\frac{v}{y} = \frac{}{y^2}$ **12.** $\frac{w}{4} = \frac{}{20}$

13. $\frac{1}{6} = \frac{2}{}$ **14.** $\frac{1}{a} = \frac{3}{}$ **15.** $\frac{1}{b} = \frac{c}{}$ **16.** $\frac{1}{d} = \frac{d}{}$

17. $\frac{3}{7} = \frac{12}{}$ **18.** $\frac{e}{f} = \frac{eg}{}$ **19.** $\frac{h}{m} = \frac{6h}{}$ **20.** $\frac{p}{t} = \frac{p^2}{}$

21. $\frac{m}{p} = \frac{}{3pq}$ **22.** $\frac{t}{u} = \frac{}{5u^2}$ **23.** $\frac{w}{7} = \frac{}{21x}$ **24.** $\frac{pt}{y^2} = \frac{}{4y^3}$.

Reduce the following fractions to their lowest terms:

25. $\frac{10}{35}$ **26.** $\frac{ax}{ay}$ **27.** $\frac{3b}{3c}$ **28.** $\frac{df}{ef}$

29. $\frac{gh}{hk}$ **30.** $\frac{n^2}{np}$ **31.** $\frac{rt}{t^2}$ **32.** $\frac{5u}{25}$

33. $\frac{7}{21}$ **34.** $\frac{b}{ab}$ **35.** $\frac{c}{c^2}$ **36.** $\frac{d^2}{d^5}$

37. $\frac{12f}{9}$ **38.** $\frac{gh^2}{g^2h}$ **39.** $\frac{5km}{10mp}$ **40.** $\frac{16n^2}{12n^3}$

41. $\frac{4q^3}{q^4}$ **42.** $\frac{6x^3}{2x^5}$ **43.** $\frac{12vy^2}{8v^2y}$ **44.** $\frac{8p^2q}{2pqx}$.

Copy and complete:

45. $5 = \frac{}{3}$ **46.** $a = \frac{}{b}$ **47.** $c = \frac{}{5}$ **48.** $4 = \frac{}{d}$

49. $1 = \frac{}{7}$ **50.** $1 = \frac{}{f}$ **51.** $g = \frac{}{g}$ **52.** $h = \frac{}{3k}$

53. $\frac{24}{3} =$ **54.** $\frac{7n}{n} =$ **55.** $\frac{2p}{2} =$ **56.** $\frac{u^3}{u} =$

57. $\frac{3}{24} = \frac{}{}$ **58.** $\frac{v}{5v} = \frac{}{}$ **59.** $\frac{8}{8x} = \frac{}{}$ **60.** $\frac{y}{y^2} = \frac{}{}$.

Reduce the following fractions to their lowest terms:

61. $\frac{a(b+c)}{d(b+c)}$ **62.** $\frac{f(g+h)}{f(k+m)}$ **63.** $\frac{10(p-q)}{15(p-q)}$ **64.** $\frac{t(x+y)}{w(x+y)}$

65. $\frac{3d+3f}{5d+5f}$ **66.** $\frac{4k-4m}{3k-3m}$ **67.** $\frac{p^2-2p}{5p-10}$ **68.** $\frac{3x-12}{x^2-4x}$

69. $\frac{6(u-3)}{(u-3)}$ **70.** $\frac{5p-5q}{p-q}$ **71.** $\frac{k(x+y)}{7k}$ **72.** $\frac{2a-2b}{6}$.

Copy and complete:

73. $\frac{h}{k} = \frac{}{k(m+n)}$ **74.** $\frac{3}{7} = \frac{}{7(p-q)}$ **75.** $\frac{u}{(x+y)} = \frac{}{p(x+y)}$

76. $\frac{v}{a+b} = \frac{}{3(a+b)} = \frac{}{3a+3b}$ **77.** $\frac{5}{8} = \frac{5(\quad)}{8(\quad)} = \frac{}{8c+8d}$

78. $\frac{m}{f+g} = \frac{}{pf+pg}$ **79.** $\frac{r}{t} = \frac{}{t^2}$

80. $\frac{u}{v+w} = \frac{}{(v+w)^2}$ **81.** $\frac{k+n}{m-p} = \frac{}{xm-xp}$.

ADDITION AND SUBTRACTION OF FRACTIONS

$$\frac{3}{5} + \frac{2}{7} = \frac{21}{35} + \frac{10}{35} = \frac{21+10}{35} = \frac{31}{35}$$

In the same way,

$$\frac{a}{b} + \frac{c}{d} = \frac{ad}{bd} + \frac{bc}{bd} = \frac{ad + bc}{bd}$$

and $$\frac{1}{3f} - \frac{2}{5g} = \frac{5g}{15fg} - \frac{6f}{15fg} = \frac{5g - 6f}{15fg}$$

Exercise 13

Simplify, giving each as a single fraction:

1. $\frac{4}{7} + \frac{1}{3}$ **2.** $\frac{w}{5} + \frac{w}{2}$ **3.** $\frac{x}{3} + \frac{y}{4}$ **4.** $\frac{a}{2} - \frac{a}{5}$

5. $\frac{p}{x} + \frac{n}{y}$ **6.** $\frac{1}{2h} + \frac{1}{3h}$ **7.** $\frac{1}{4m} - \frac{1}{7m}$ **8.** $\frac{2}{5y} - \frac{3}{8y}$

9. $\frac{c}{3a} + \frac{d}{2b}$ **10.** $\frac{4}{5g} + \frac{2}{3f}$ **11.** $\frac{2}{9} + \frac{1}{3}$ **12.** $\frac{5}{k^2} + \frac{1}{k}$

13. $\frac{2}{m} - \frac{4}{m^2}$ **14.** $\frac{5}{3p} - \frac{2}{p^2}$ **15.** $\frac{7}{10} - \frac{4}{15}$ **16.** $\frac{x}{uv} - \frac{y}{vw}$.

$$2\frac{3}{5} = 2 + \frac{3}{5} = \frac{10}{5} + \frac{3}{5} = \frac{13}{5}$$

Similarly $$2 + \frac{a}{b} = \frac{2b}{b} + \frac{a}{b} + \frac{2b + a}{b}$$

Copy and complete:

17. $3\frac{4}{7} = \frac{\ldots}{7}$ **18.** $2 + \frac{f}{3} = \frac{\ldots + f}{3}$

19. $5 + \frac{d}{c} = \frac{\ldots + \ldots}{c}$ **20.** $h + \frac{5}{k} = \frac{\ldots + \ldots}{k}$

21. $\frac{7}{m} + p = \frac{\ldots + \ldots}{m}$ **22.** $6 - \frac{t}{q} = \frac{\ldots - \ldots}{q}$

23. $\frac{v}{x} - y = \frac{\qquad}{x}$ **24.** $3y - \frac{3}{y} = \frac{\qquad}{y}$

25. $\dfrac{4c + 3d}{12} = \dfrac{4c}{12} + \dfrac{\dots}{12} = \dfrac{c}{3} + \dfrac{\dots}{4}$

26. $\dfrac{5a + 5b}{ab} = \dfrac{\dots}{ab} + \dfrac{\dots}{ab} = \dfrac{\dots}{b} + \dfrac{\dots}{a}$

Use the method of Questions **25** and **26** to express the following as two fractions added together:

27. $\dfrac{3x + 5y}{15}$ **28.** $\dfrac{py + xt}{xy}$ **29.** $\dfrac{2bc + 3ad}{6ab}$ **30.** $\dfrac{5 + k}{k^2}$

Express as the difference of two fractions in their lowest terms:

31. $\dfrac{b - c}{bc}$ **32.** $\dfrac{7h - 3m}{21}$ **33.** $\dfrac{n - 4}{4n}$ **34.** $\dfrac{4v - 2p}{8pv}$

MULTIPLICATION OF FRACTIONS

EXAMPLES: $\dfrac{2}{3} \times \dfrac{9}{10} = \dfrac{18}{30} = \dfrac{3}{5}$

$\dfrac{a}{b} \times \dfrac{bc}{a^2} = \dfrac{abc}{ba^2} = \dfrac{ac}{a^2} = \dfrac{c}{a}$

Exercise 14

Simplify:

1. $\dfrac{3}{5} \times \dfrac{2}{7}$ **2.** $\dfrac{c}{x} \times \dfrac{b}{y}$ **3.** $\dfrac{2}{a} \times \dfrac{b}{3}$ **4.** $\dfrac{f}{g} \times \dfrac{f}{h}$

5. $\dfrac{2}{3} \times \dfrac{9}{8}$ **6.** $\dfrac{a}{6} \times \dfrac{3}{b}$ **7.** $\dfrac{d}{f} \times \dfrac{h}{d}$ **8.** $\dfrac{g}{m} \times \dfrac{m^2}{gk}$

9. $\dfrac{6}{7} \times \dfrac{5}{6}$ **10.** $\dfrac{n}{7} \times \dfrac{5}{n}$ **11.** $\dfrac{(a + b)}{7} \times \dfrac{5}{(a + b)}$

12. $\dfrac{2}{(c - d)} \times \dfrac{(c - d)}{3}$ **13.** $\dfrac{3}{4} \times \dfrac{4}{3}$ **14.** $\dfrac{a}{b} \times \dfrac{b}{a}$

15. $\dfrac{p}{x} \times \dfrac{x}{y}$ **16.** $\dfrac{n}{6} \times \dfrac{4}{n}$ **17.** $\dfrac{2}{5} \times \dfrac{5}{6}$ **18.** $\dfrac{y}{3} \times \dfrac{3}{y^2}$

19. $\dfrac{a^2}{5} \times \dfrac{10}{a}$ **20.** $\dfrac{2b}{c^2} \times \dfrac{c}{b^2}$ **21.** $\dfrac{h}{(p + q)} \times \dfrac{(p + q)}{n}$

22. $\frac{5(m+g)}{2} \times \frac{1}{(m+g)}$ **23.** $\frac{4(p+t)}{5} \times \frac{1}{2(p+t)}$

24. $\frac{9}{(x-y)v} \times \frac{(x-y)w}{3}$

25. Copy and complete:

$$\frac{(4x+4y)}{(wx+wy)} \times \frac{w}{6} = \frac{4(\ldots+\ldots)}{w(\ldots+\ldots)} \times \frac{w}{\ldots} = \frac{4}{\ldots} \times \frac{\ldots}{\ldots} = \frac{\ldots}{\ldots}$$

Simplify as in Question **25**:

26. $\frac{(u^2+u)}{(2u+2)} \times \frac{3}{u}$ **27.** $\frac{(px+py)}{5} \times \frac{1}{(qx+qy)}$

28. $\frac{a^2}{(3b-3c)} \times \frac{(b-c)}{4a}$ **29.** $\frac{2h}{(m-5)} \times \frac{(hm-5h)}{6h^2}$

DIVISION OF FRACTIONS

$$\frac{3}{4} \div \frac{2}{5} = \frac{15}{20} \div \frac{8}{20} = \frac{15 \text{ twentieths}}{8 \text{ twentieths}} = \frac{15}{8}$$

Notice that $\frac{3}{4} \times \frac{5}{2} = \frac{15}{8}$ also.

Now examine the following:

$$\frac{a}{b} \div \frac{c}{d} = \frac{ad}{bd} \div \frac{bc}{bd} = \frac{\frac{ad}{bd}}{\frac{bc}{bd}} = \frac{ad}{bc} = \frac{a}{b} \times \frac{d}{c}$$

Dividing by $\frac{c}{d}$ has the same result as multiplying by $\frac{d}{c}$.

This gives the following working rule:

To divide by a fraction, multiply by the fraction inverted.

EXAMPLES:
$$\frac{2}{7} \div \frac{5}{9} = \frac{2}{7} \times \frac{9}{5} = \frac{18}{35}$$
$$\frac{5}{6} \div \frac{1}{3} = \frac{5}{6} \times \frac{3}{1} = \frac{15}{6} = \frac{5}{2} = 2\frac{1}{2}$$
$$\frac{a^2}{b} \div \frac{ac}{b} = \frac{a^2}{b} \times \frac{b}{ac} = \frac{a^2b}{abc} = \frac{a^2}{ac} = \frac{a}{c}$$

Exercise 15

Simplify:

1. $\frac{2}{5} \div \frac{3}{7}$ **2.** $\frac{3}{7} \div \frac{2}{5}$ **3.** $\frac{p}{r} \div \frac{u}{w}$ **4.** $\frac{u}{w} \div \frac{p}{r}$

5. $\frac{5}{7} \div \frac{3}{4}$ **6.** $\frac{2}{3} \div \frac{3}{1}$ **7.** $\frac{1}{5} \div \frac{2}{7}$ **8.** $\frac{1}{4} \div \frac{2}{9}$

9. $\frac{a}{3} \div \frac{b}{4}$ **10.** $\frac{h}{2} \div \frac{5}{m}$ **11.** $\frac{n}{2} \div \frac{3}{n}$ **12.** $\frac{1}{p} \div \frac{p}{7}$

13. $\frac{2}{7} \div \frac{2}{3}$ **14.** $\frac{a}{5} \div \frac{a}{2}$ **15.** $\frac{x}{6} \div \frac{y}{9}$ **16.** $\frac{p}{n} \div \frac{q}{kn}$

17. $\frac{vt}{w} \div \frac{v^2}{w}$ **18.** $\frac{x^2}{y^2} \div \frac{x}{y}$ **19.** $\frac{(a+b)}{5} \div \frac{(a+b)}{3}$

20. $\frac{4}{(c-d)} \div \frac{6}{(c-d)}$ **21.** $\frac{2(e+f)}{5} \div \frac{4(e+f)}{15}$

22. $\frac{(3g+3h)}{2} \div \frac{(2g+2h)}{3}$ **23.** $\frac{1}{(px+py)} \div \frac{2}{(qx+qy)}$

EQUATIONS

EXAMPLE: *Four boys arranged to hire a rowing boat and to share the cost equally. On the day, one of them did not arrive and the cost was shared equally by the other three. Each had to pay 5p more than he expected. Find the cost of hiring the boat.*

Let the cost be xp.

Then each boy expected to pay $\frac{x}{4}$p.

The three who used the boat each paid $\frac{x}{3}$p.

We know that $\frac{x}{3}$p is 5p more than $\frac{x}{4}$p.

This statement gives the following equation:

$$\frac{x}{3} = \frac{x}{4} + 5$$

Multiplying both sides by 12,

$$\frac{12\,x}{3} = \frac{12\,x}{4} + 60$$

$$4\,x = 3x + 60$$

$$x = 60$$

The hire charge was 60p.

When solving an equation containing fractions, multiply by the smallest number which will remove the fractions. For $\frac{x}{9} + \frac{x}{6} = 5$, should you multiply by 54, 18, 9 or 6?

Exercise 16

Solve the following equations:

1. $\frac{x}{3} + \frac{x}{2} = 10$ **2.** $\frac{x}{2} - \frac{x}{5} = 9$ **3.** $\frac{2x}{3} + \frac{3x}{4} = 34$

4. $\frac{x}{3} - \frac{x}{5} = \frac{4}{3}$ **5.** $\frac{2x}{5} - \frac{3x}{7} + \frac{4}{7} = 0$ **6.** $\frac{x}{3} + \frac{x}{6} + 4 = x$

7. $\frac{x}{4} - \frac{x}{8} = 5$ **8.** $\frac{5x}{6} - \frac{3x}{4} = \frac{2}{3}$ **9.** $\frac{7x}{10} = \frac{2x}{5} + 6$

10. $\frac{x}{4} = \frac{x}{3} - \frac{1}{6}$ **11.** $\frac{x-3}{5} = \frac{x}{2}$ **12.** $\frac{x+1}{3} + \frac{x-1}{4} = \frac{1}{2}$

13. $\frac{x+2}{5} = \frac{x+3}{4}$ **14.** $\frac{x-3}{2} = \frac{x+1}{4} + \frac{x-5}{8}$

15. At the start of a game, x counters were shared equally amongst 6 players. How many did each get? The losing player dropped out

and in the next game the x counters were shared amongst the remaining 5 players. Each got 2 more than in the first game. Set up an equation for x and solve it.
In the third game the counters were shared amongst 4 players. How many did each get?

16. Five boys shared a prize of £y. How much did each get? Later three girls shared a prize of the same value. Each girl received £4 more than each boy. Set up an equation for y and solve it.

17. When some cards are dealt to 8 people each receives 3 fewer than when they are dealt to 6 people. How many cards are there?

18. In ΔABC, $\hat{B}$ is $\frac{2}{3}$ of $\hat{A}$ and $\hat{C}$ is $\frac{5}{6}$ of $\hat{A}$. Let $\hat{A} = x°$. Write down the values of $\hat{B}$ and $\hat{C}$. Form an equation using the angle sum of a triangle. Solve it and state the sizes of the three angles.

19. John takes 15 paces across a classroom and Pete takes 20 paces. If each of John's paces is 20 cm longer than each of Pete's, find the width of the room.

5 · FORMULAE

Mary is saving up to buy Christmas presents. She now has 250 pence in her cash box. If she puts 40 pence in the box each week for 9 weeks she will have

$$250 + 40 \times 9 = 610 \text{ pence.}$$

If she puts 40 pence in the box for n weeks she will have

$$250 + 40n \text{ pence}$$

Call this c pence, then

$$c = 250 + 40n$$

This is a formula for the amount of cash in the box after any number of weeks. For example, if $n = 5$, then $c = 250 + 40 \times 5 = 250 + 200 = 450$ and so after 5 weeks there are 450 pence in the box.

Now suppose that Mary has b pence in her box and puts in p pence each week. Then after n weeks she has c pence where

$$c = b + pn$$

This is a more general formula.

If $b = 120$, $p = 70$ and $n = 8$,
then $c = 120 + 70 \times 8 = 120 + 560 = 680$.

An aircraft flies for 3 hours at 620 km/h.
It travels $620 \times 3 = 1860$ km.
If a ship travels at v km/h for t h and covers a distance of d km, then

$$d = vt$$

This is a formula for the distance in terms of the speed and time.
If $v = 18$ and $t = 4$, then $d = 18 \times 4 = 72$. This means that if the speed is 18 km/h and the time is 4 h, then the distance travelled is 72 km.

Exercise 17

1. A boy is now 10 years old. In n years time he will be y years old. Write down a formula for y.

2. I face North and turn clockwise through x°, where x is less than 90. In order to face East I must turn a further y°. Write down a formula for y.

3. (i) An aircraft has to make a journey of 2400 kilometres. Its speed is 600 km/h. How long does it take?
   (ii) Write down a formula for the time, t hours, taken to travel d km at a speed of s km/h.
   (iii) Use the formula to find the time taken by a ship to travel 70 km at a speed of 20 km/h.

4. A lorry weighs x tonnes when empty. When it has a load of y tonnes of sand, it weighs p tonnes. Write down a formula for y in terms of x and p. On a weighbridge the weight of a lorry is shown as 12.3 tonnes. The lorry weighs 4.9 tonnes when empty. Use the formula to find the load.

5. A city reservoir contains k litres of water. Each day x litres are drawn off. After n days in which there has been no rain, p litres are left. Write down a formula for p.
   If $k = 2300$ million, $x = 60$ million and $n = 15$, find p.

6. A girl lives x kilometres from her school and cycles to school and back home each day. Her cyclometer records p kilometres one evening and after n more schooldays it records q kilometres. Write down a formula for q. (The girl has dinner at school and always uses the same route.)

7. The cost of running a coach on a day trip to the seaside is £x. The coach carries n passengers who each pay y pence. Write down a formula for the profit £p which the coach company makes.
   Use your formula to find the profit for a coach containing 30 passengers who pay 160 pence each if the coach costs £34 to run. What happens if there are only 15 passengers?

FORMULAE FROM STATEMENTS

Here is a rule given in a cookery book:

To find the time needed to cook a piece of meat you should allow 30 minutes for each kilogramme and an extra 20 minutes.

We can obtain a formula from this statement.

Let the time for a mass of m kilogrammes be t minutes, then

$$t = 30m + 20$$

Exercise 18

Write down a formula from each of the following statements.

1. The length of a rectangle can be found by dividing its area by its width. (Use L, A and W.)

2. The perimeter of a square is four times the length of a side. (Use p and x.)

3. The average speed for a journey is found by dividing the distance travelled by the time taken. (Use s, d and t.)

4. The area of a label needed for a cylindrical tin is found by multiplying together π, the diameter and the height. (Use A, d, π and h.)

5. From a cliff top, the distance of the horizon in kilometres is found by multiplying the height of the cliff in metres by 13 and then taking the square root. (Use d and h.)

6. The volume of a sphere is 4.2 times the cube of the radius. (Use V and r.)

7. Given one of the base angles of an isosceles triangle, the angle at the apex can be found by doubling this base angle and subtracting the answer from 180°.

8. The perimeter of a rectangle is found by adding together the length and breadth and multiplying the answer by 2.

9. The distance in kilometres of a flash of lightning can be estimated by counting the number of seconds between seeing the flash and hearing the thunder clap and dividing this number by 3.

10. To find the depth of a well, drop a stone into it and count the number of seconds before it hits the bottom. The depth in metres is 4.9 times the square of this number of seconds.

The following formulae appeared in Books 1 and 2. Explain them:

11. $A = \frac{1}{2}bh$ for a triangle

12. $V = lbh$

13. $c = \pi d$ for a circle

14. $A = \pi r^2$ for a circle

15. $a + b + c = 180$ for a triangle

16. $a^2 = b^2 + c^2$ for a right-angled triangle

17. $s = (n - 2) \times 180$ for a polygon

REARRANGING FORMULAE

A boy has £3 in a cash box and decides to add 20p each week. After n weeks he has a sum of £C where $C = \frac{1}{5}n + 3$.

He wonders whether to save up for a radio costing £14, a bicycle costing £18 or a guitar costing £8 and he wants to know how long it would take for each. As he requires the value of n for three different values of C, it is helpful to rearrange the formula so that it reads $n = \ldots$.

We say that we are making n the *subject* of the formula.

We start with $C = \frac{1}{5}n + 3$
Subtracting 3 from both sides, $C - 3 = \frac{1}{5}n$.
Multiplying both sides by 5, $5(C - 3) = n$.

The formula is $n = 5(C - 3)$
or $n = 5C - 15$

When $C = 14$, $n = 5 \times 14 - 15 = 70 - 15 = 55$
When $C = 18$, $n = 5 \times 18 - 15 = 90 - 15 = 75$
When $C = 8$, $n = 5 \times 8 - 15 = 40 - 15 = 25$

The steps in rearranging a formula are the same as the steps used in solving an equation. It is important to understand these steps fully. The next exercise starts with questions designed to revise equation work.

Exercise 19

Solve the following equations. At each step state what you are doing to the two sides of the equation.

1. $x + 5 = 9$ **2.** $5x = 9$ **3.** $\frac{1}{5}x = 9$

4. $x - 5 = 9$ **5.** $x^2 = 9$ **6.** $2x^2 = 50$

7. $\sqrt{x} = 9$ **8.** $3\sqrt{x} = 18$ **9.** $\frac{5}{x} = 9$

10. $\frac{2}{3}x = 5$ **11.** $\frac{x}{7} = \frac{3}{4}$ **12.** $\frac{21}{x} = \frac{2}{5}$

13. $4x + 7 = 25$ **14.** $\frac{1}{5}x - 3 = 4$

Make x the subject of each of the following formulae:

15. $x + a = b$ **16.** $cx = d$ **17.** $\frac{x}{f} = g$

18. $x - h = k$ **19.** $x^2 = n$ **20.** $px^2 = q$

21. $\sqrt{x} = p$ **22.** $v\sqrt{x} = y$ **23.** $\frac{r}{x} = t$

24. $\frac{u}{w}x = y$ **25.** $\frac{x}{a} = \frac{b}{c}$ **26.** $\frac{d}{x} = \frac{e}{f}$

27. $gx + h = k$ **28.** $\frac{x}{n} - k = p$

Rearrange each of the following formulae so that the stated letter is the subject:

29. $c = \pi d;\ d$ **30.** $V = lbh;\ h$
31. $a + b + c = 180;\ b$ **32.** $A = \frac{1}{2}bh;\ h$
33. $s = \frac{d}{t};\ d$ **34.** $s = \frac{d}{t};\ t$
35. $A = \pi r^2;\ r$ **36.** $d = \sqrt{(13h)}\ ;\ h$
37. $s = 180(n - 2);\ n$ **38.** $a^2 = b^2 + c^2;\ b$
39. $v = u + at;\ a$ **40.** $t = \frac{1}{3}\sqrt{s};\ s$

Exercise 20

1. When its speed is k kilometres per hour a car travels m metres each second, where $m = \frac{5}{18}k$.

(i) How far does it travel each second if its speed is
(a) 54 km/h (b) 72 km/h?

(ii) Make k the subject of the formula.

(iii) What is the speed in km/h if the distance travelled in a second is
(a) 25 m (b) 42.5 m?

2. A mathematician gives his son w pence pocket money each week where w is calculated from the formula $w = 20g - 180$, g being the son's age in years. ($g > 9$)

(i) Calculate the pocket money the boy gets when he is 12 years old.

(ii) Rearrange the formula to make g the subject.

(iii) At what age does the boy get:
(a) 80p per week (b) £1.20 per week?

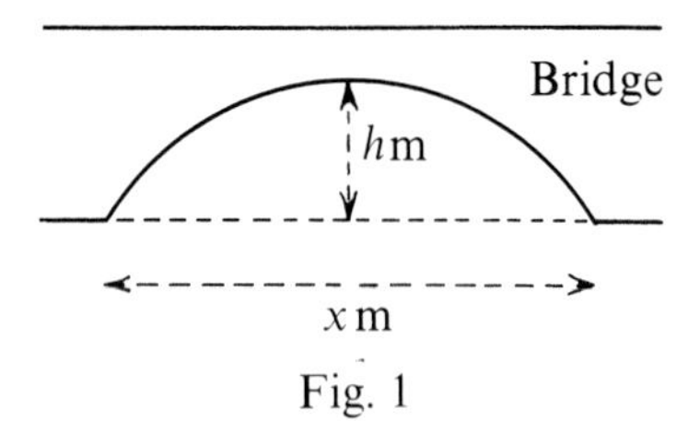

Fig. 1

3. The figure shows the arch of a bridge. The formula for x in terms of h is $x = \sqrt{(200h)}$.

(i) Find x if $h = 8$.
(ii) Rearrange the formula to make h the subject.
(iii) Calculate h if $x = 60$.

4. Write down a formula for the time, t hours, taken to travel at v km/h from London to Edinburgh, a distance of 600 km.

(i) Find t for $v = 75$ and for $v = 120$.
(ii) Make v the subject of the formula.
(iii) What average speed must the motorist maintain in order to complete the journey in $7\frac{1}{2}$ hours?

5. In $\triangle ABC$, $\hat{B} = \hat{C} = x°$ and $\hat{A} = y°$.

(i) Write down a formula expressing y in terms of x.
(ii) If $x = 62$, calculate y.
(iii) Rearrange the formula to make x the subject.
(iv) If $y = 106$, calculate x.

6. A rectangular box has a square base of side x cm and a height of 16 m. The volume is v cm^3.

(i) Write down the formula for v in terms of x.
(ii) Rearrange it to make x the subject.
(iii) Calculate x if $v = 400$.

7. The size $x°$ of an angle of a regular polygon of n sides is given by the formula $x = 180 - \dfrac{360}{n}$.

(i) Calculate the angle for a pentagon and for a 10-sided polygon.
(ii) Rearrange the formula to make n the subject.
(iii) How many sides has a regular polygon if its angle is:
(a) 140° (b) 160° (c) 156°?
(iv) What happens if x is taken as 110°? Explain.

8. A box of counters is passed round a class. After the box has been to p pupils there are n counters in it where $n = 200 - 6p$.

(i) Explain the formula.
(ii) Find n when $p = 12$.

(iii) State the maximum possible value of p.
(iv) Rearrange the formula so that p is the subject.
(v) Find p when $n = 20$.

9. The cost of n copies of a certain textbook is £C where $C = \dfrac{xn}{100}$.

(i) Explain the meaning of x.
(ii) Find the value of C if $x = 75$ and $n = 16$.
(iii) Rearrange the formula to make n the subject.
(iv) Find n if $C = 12$ and $x = 60$.

10. The weekly cost of running a small guest house is given by the formula $C = 360 + 10n$.

(i) Explain the formula.
(ii) Find C if $n = 16$.
(iii) Rearrange the formula so that n is the subject.
(iv) Find n if $C = 580$.
(v) If each guest pays £30 write down the total sum received from n guests. Put this equal to C and solve the equation obtained. How many guests are needed in order to make a profit?

11. A ladder of length 10 m has its foot d m from a vertical wall and reaches a point h m above the ground.

(i) Derive a formula for h in terms of d.
(ii) Calculate h if d is (a) 6 (b) 3.

6 · SQUARES AND SQUARE ROOTS

$80^2 = 80 \times 80 = 6400$ and so $\sqrt{6400}$ is 80

$0.2^2 = 0.2 \times 0.2 = 0.04$ and so $\sqrt{0.04}$ is 0.2

$3.4^2 = 3.4 \times 3.4 = 11.56$ and so $\sqrt{11.56}$ is 3.4

$$\sqrt{2\frac{7}{9}} = \sqrt{\frac{25}{9}} = \frac{\sqrt{25}}{\sqrt{9}} = \frac{5}{3} = 1\frac{2}{3}$$

Exercise 21

1. Work out the squares of:

(i) 7 (ii) 70 (iii) 16 (iv) 23
(v) 2.3 (vi) 230 (vii) 5.8 (viii) 4.2.

2. Work out the squares of:

(i) 0.3 (ii) 0.9 (iii) 0.09 (iv) 0.01
(v) 0.12 (vi) 1.2 (vii) 120 (viii) 1200.

3. State the square root of:

(i) 36 (ii) 900 (iii) 0.36 (iv) 0.09
(v) 1600 (vi) 1.44 (vii) 400 (viii) 0.04.

4. State the length of a side of a square which has an area of:

(i) $100\ \text{m}^2$ (ii) $2500\ \text{m}^2$.

5. Calculate 6.7^2. What is $\sqrt{44.89}$

6. Calculate 5.3^2. If $\sqrt{n} = 5.3$, what is the value of n?

7. Calculate 340^2. If $\sqrt{k} = 340$, what is the value of k?

8. Work out the squares of $\frac{2}{3}, \frac{1}{4}, \frac{3}{7}, 1\frac{1}{2}, 3\frac{1}{2}$.

9. Find out the square roots of $\frac{9}{16}, \frac{1}{9}, 2\frac{1}{4}, 1\frac{7}{9}, 6\frac{1}{4}$.

10. A square has a side of 6.4 cm. Find its area, correct to 2 sig. fig.

11. Calculate 3.55^2 and give your answer correct to 3 sig. fig.

12. Calculate 23.4^2 and give your answer correct to 3 sig. fig.

USING TABLES OF SQUARES

$$6.37^2 = 40.5769$$

Correct to 4 sig. fig. this is 40.58

There are tables which give the squares of numbers correct to 4 sig. fig. The diagram shows how to use such a table to find the square of 6.37 (correct to 4 sig. fig.)

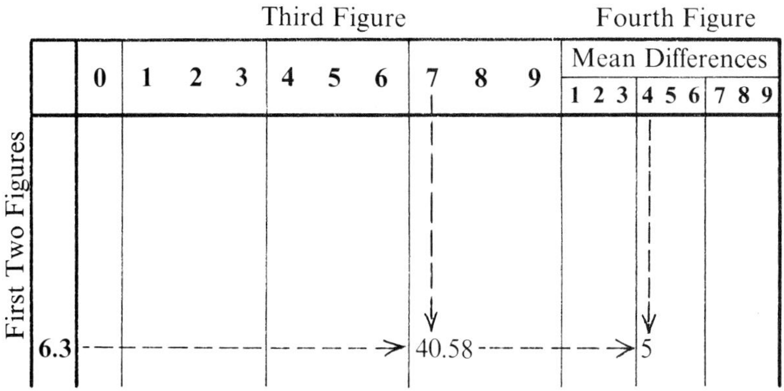

	Third Figure										Fourth Figure: Mean Differences								
First Two Figures	0	1	2	3	4	5	6	7	8	9	1	2	3	4	5	6	7	8	9
6.3								40.58						5					

For 6.374^2, we use the 'Mean Differences' column headed 4. The number on the 6.3 row is 5. This is added to the last figure of 40·58 giving 40·63. (The exact square of 6.374 is 40.627 876.)

The tables give the squares of numbers from 1.000 to 9.999. For larger or smaller numbers we can work as follows:

$$63.7^2 = (6.37 \times 10)^2 = 6.37^2 \times 10^2$$
$$\simeq 40.58 \times 100 = 4058$$
$$(0.637)^2 = (6.37 \times 0.1)^2 = 6.37^2 \times 0.1^2$$
$$\simeq 40{\cdot}58 \times 0.01 = 0.4058$$

Alternatively we can use a rough estimate:

As $60^2 = 3600$, $\quad 63.7^2 \simeq 4058$

As $0.6^2 = 0.36$, $\quad 0.637^2 \simeq 0.4058$

Exercise 22

Use tables to obtain approximate squares of:

1. 2.73, 1.56, 6.27, 7.34.

2. 3.5 (look up 3.50), 4.9, 8.4, 5.8.

3. 1.473, 2.542, 4.436, 5.774.

4. 2.96, 29.6, 0.296.

5. 1.53, 153, 0.0153.

6. 8.39, 0.839, 83.9.

7. 7.58, 758, 0.758.

8. 3.523, 35.23, 352.3.

9. 6.072, 0.6072, 0.060 72.

10. 8.355, 83.55, 0.083 55.

USING TABLES OF SQUARE ROOTS

In Book 2 we found that some numbers have exact square roots, but others have not.

$\sqrt{25}$ is 5 and $\sqrt{324}$ is 18 but there is no exact square root for 29 or for 467.

$5.38^2 = 28.9444$ and $5.39^2 = 29.0521$ so that $\sqrt{29}$ is between 5.38 and 5.39.

5.385^2 is less than 29 but 5.386^2 is greater than 29 so that $\sqrt{29}$ is between 5.385 and 5.386. It is nearer to 5.385 and so we can say that $\sqrt{29}$ is 5.385, correct to 4 sig. fig.

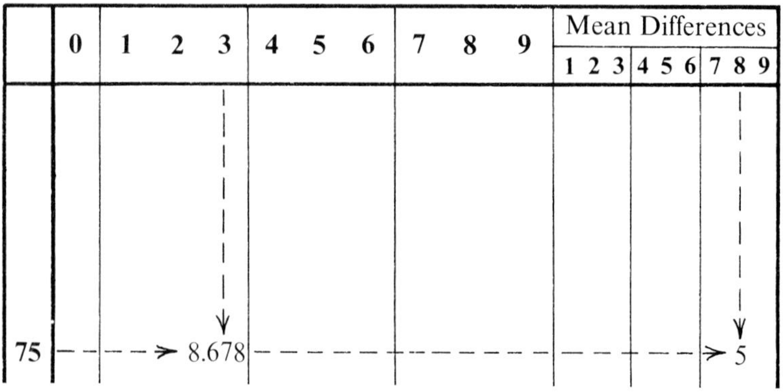

	0	1	2	3	4	5	6	7	8	9	Mean Differences								
											1	2	3	4	5	6	7	8	9
75				8.678														5	

There are tables which give the square roots, correct to 4 sig. fig., of numbers from 1 to 100.

The diagram shows how to use a table to find the square root of 75.38

$$\sqrt{75.3} \simeq 8.678$$
$$\sqrt{75.38} \simeq 8.678 + 0.005 = 8.683$$

For numbers greater than 100 or less than 1, either of the following methods can be used:

METHOD 1: As $592 = 5.92 \times 100$

$$\sqrt{592} = \sqrt{5.92} \times \sqrt{100}$$
$$\simeq 2.433 \times 10 = 24.33$$

(Check: $20 \times 20 = 400$)

As $0.008\,77 = 87.7 \div 10\,000$

$$\sqrt{0.008\,77} = \sqrt{87.7} \div \sqrt{10\,000}$$
$$\simeq 9.365 \div 100 = 0.093\,65$$

(Check: $0.09 \times 0.09 = 0.0081$)

METHOD 2: To find $\sqrt{32\,864}$

Divide the digits into pairs from the right: 3'28'64'

Look up $\sqrt{3.286}$ obtaining 1.813

Write $\sqrt{3'28'64}$

$\simeq$ 1 8 1 . 3 so that there is one figure in the answer below each pair of digits in the given number.

To find $\sqrt{0.008\,77}$

Dividing into pairs from the decimal point, we have

0.'00'87'7.

Look up $\sqrt{87.7}$ obtaining 9.365

Write $\sqrt{0.'0\ 0'8\ 7'7}$

$\simeq$ 0. 0 9 3 6 5 so that there is one figure in the answer below each pair in the given number.

Exercise 23

Use tables to obtain approximate square roots of:

1. 44, 69, 4.4, 6.9.

2. 2.58, 6.97, 13.6, 83.9.

3. 4.03, 7.56, 23.7, 68.7.

4. 3.524, 5.008, 62.44, 43.09.

5. 263, 895, 6320, 7650.

6. 3238, 568, 8214, 9716.

7. 9, 90, 900, 9000.

8. 2.5, 25, 250, 2500.

9. 0.14, 0.68, 0.5, 0.7.

10. 0.0636, 0.0157, 0.872, 0.932.

11. 0.283, 0.0283, 0.002 83, 0.000 283.

12. 0.0694, 0.694, 0·000 069 4, 0·000 694.

13. 472.7, 47.27, 4.727, 0.472 7.

14. 85.38, 853.8, 0.085 38, 0.8538.

15. From the list of numbers below, pick out the approximate square roots of:
(i) 12 (ii) 1.2 (iii) 45 (iv) 258 (v) 0.37.
List: 0.062, 0.19, 0.62, 1.1, 2.1, 3.5, 6.7, 16, 51.

16. Find the approximate length of the side of a square of area 57 cm^2.

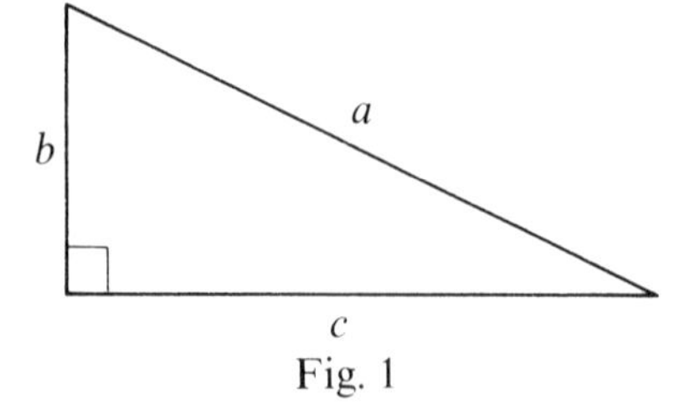

Fig. 1

Pythagoras' Theorem:
$a^2 = b^2 + c^2$

17. If $b = 7$ and $c = 9$, calculate a to 2 sig. fig.

18. If $b = 3.6$ and $c = 6.7$, calculate a to 2 sig. fig.

19. If $a = 10$ and $c = 7$, calculate b to 2 sig. fig.

20. If $a = 42$ and $b = 35$, calculate c to 2 sig. fig.

21. A ladder of length 6 m is placed on horizontal ground with its foot 3 m from a vertical wall. How far up the wall does it reach?

22. A rectangle has sides of 9 cm and 13 cm. Calculate the length of a diagonal.

23. Make a sketch showing axes Ox and Oy and the points P as (3, 7) and Q as (9, 4). Calculate the lengths of OP, OQ and PQ.

24. A flag-pole 10 m high is to be held by three guy-ropes attached to the pole at 7 m above the ground and to points on the ground 4 m from the pole. Calculate the length of each rope.

PYTHAGOREAN TRIADS

If the values $b = 2$ and $c = 3$ are substituted in the formula $a^2 = b^2 + c^2$ then $a^2 = 4 + 9 = 13$ and so $a = \sqrt{13}$, for which we can only give an approximate value, but if $b = 3$ and $c = 4$ are substituted in the formula $a^2 = 9 + 16 = 25$ and $a = 5$, an integer.

There are many sets of three integers which fit the formula $a^2 = b^2 + c^2$. They are called *Pythagorean triads.* Check that 8, 15 and 17 fit the formula. Do 48, 55 and 73?

THE CONVERSE OF PYTHAGORAS' THEOREM

Draw a triangle with sides of 85 mm, 77 mm and 36 mm. Measure the largest angle. Notice that $77^2 + 36^2 = 5929 + 1296 = 7225 = 85^2$. This illustrates the converse of Pythagoras' Theorem: *if the square of the longest side of a triangle is equal to the sum of the squares of the other two sides, the largest angle is a right-angle.*

This fact can be used to obtain a right-angle when marking out a football pitch or the foundation of a building. The ancient Egyptians used this method. They had a continuous rope loop divided into twelve equal lengths.
Posts were placed in the ground so that the loop formed a triangle with sides of 3, 4 and 5 units of length.
Try this with a piece of string or rope.

Fig. 2

Exercise 24

1. Which of the following sets of integers are Pythagorean triads?

(i) 5, 12, 13 (ii) 11, 15, 19
(iii) 20, 23, 28 (iv) 20, 21, 29
(v) 26, 43, 50 (vi) 28, 45, 53
(vii) 12, 35, 37 (viii) 23, 35, 39.

2. A football pitch has been marked out on a school field. The sides are 120 m and 90 m. To check the right-angles the two diagonals are to be measured. How long should they be?

3. Three triangles have sides of the following lengths:

$\triangle$I 3 cm, 4 cm and 5 cm; $\triangle$II 6 cm, 8 cm and 10 cm;
$\triangle$III 15 cm, 20 cm and 25 cm.

Show that sides of each triangle fit the formula $a^2 = b^2 + c^2$ and hence that each triangle has a right-angle.

Explain why the triangles are similar.

Explain how the sides of $\triangle$II and $\triangle$III can be obtained from the sides of $\triangle$I.

$\triangle$IV is similar to $\triangle$I and the length of its hypotenuse is 45 cm. State the lengths of the other two sides.

4. Fig. 3 shows two similar triangles

(i) State the length of QR.
(ii) State TS: QP.
(iii) Hence state the length of SV and of TV.

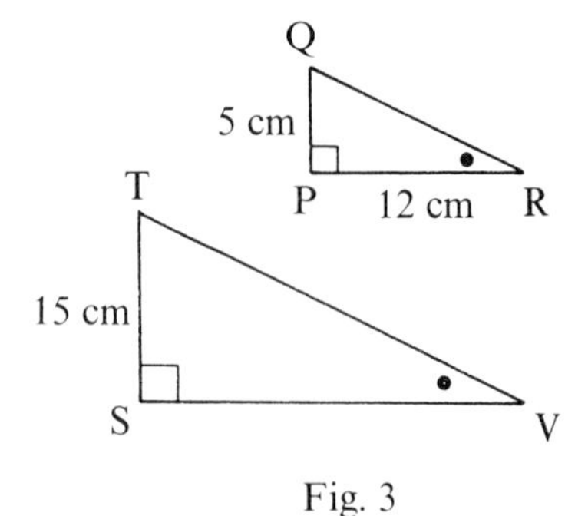

Fig. 3

5. (i) Show that a triangle with sides of 8 cm, 15 cm and 17 cm has a right-angle.

(ii) A second triangle is similar to the triangle of (i) and has a hypotenuse of length 8.5 cm. State the lengths of the other two sides.

(iii) A third triangle is similar to the others and its shortest side is 32 cm. Find the other two sides.

6. In Fig. 4, by first finding x^2 and y^2, show that one of the angles of $\triangle PQR$ is $90°$.

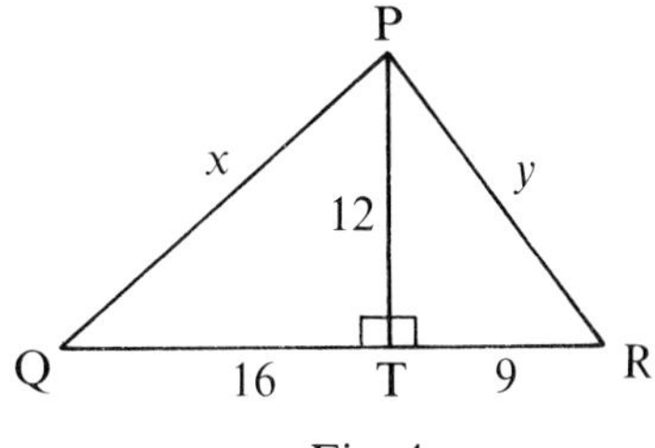

Fig. 4

7. In Fig. 5, find p and show that $\theta = 90°$.

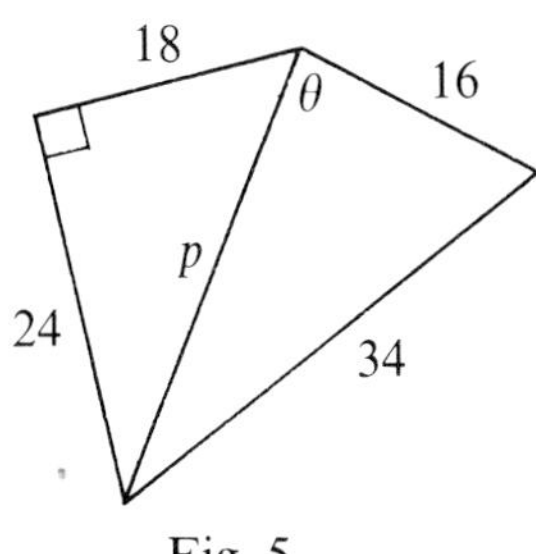

Fig. 5

8. ABCD is a square (Fig. 6). Find BP^2, PQ^2 and BQ^2.
Show that angle $PQB = 90°$.

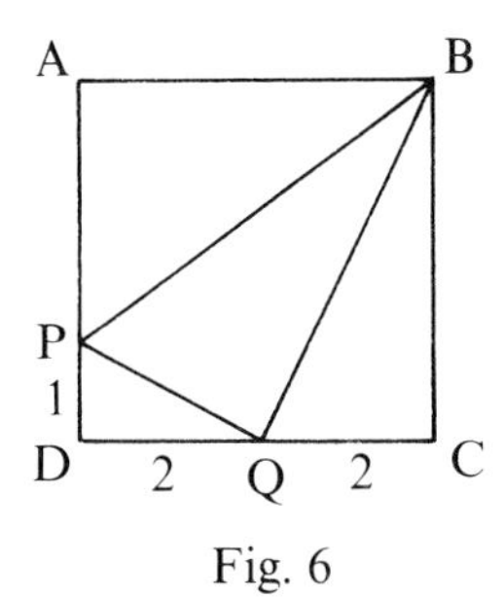

Fig. 6

9. A triangle having sides of $2p$, $p^2 - 1$ and $p^2 + 1$ units of length has a right-angle. What sides are obtained by taking p as:
(i) 2 (ii) 3 (iii) 4?

10. A triangle having sides of $2fg$, $f^2 - g^2$ and $f^2 + g^2$ units has a right-angle. What sides are obtained by taking:
(i) $f = 5$ and $g = 2$ (ii) $f = 7$ and $g = 4$?
Try some other pairs of values for f and g.

7 · PROBABILITY

The seven sentences below have spaces in them. For each space choose a word or phrase from this list: likely, impossible, unlikely, certain, 50–50 chance.

1. It is . . . to run 100 metres in 5 seconds.
2. The bus is . . . to be on time.
3. The school team has a . . . of winning the match.
4. We are . . . to have a white Christmas.
5. In tossing a coin there is a . . . of getting a head.
6. You are . . . to get an odd or even number in one throw of a dice.
7. He is . . . to win the chess tournament.

You have been using your experience and judgement to assess the probability that an event will take place. Words such as 'likely' and 'probably' are used when we expect an event to happen, but there is a chance that it may not. A phrase like '50–50 chance' suggests that success and failure are equally likely.

QUESTIONS FOR CLASS DISCUSSION

1. How can we measure the chance of a car driver having an accident?
2. What is the chance of winning a premium bond prize?
3. How does a bookmaker or pools firm use probability?

EXPERIMENTS

The following experiments will give you practical experience of probability. They need simple recordings and calculations. If groups of three work together, one can do the experiment, one record the results and one check. It is useful for a class to pool the results of each experiment and then to discuss them.

1. Throw a die 100 times. On a tally chart, record the number of sixes obtained. Also record the throws so that you know when you have reached 100.

Throws ~~||||~~ ~~||||~~ $_{10}$ ~~||||~~ ~~||||~~ $_{20}$ ~~||||~~ ||| . . .
Sixes ~~||||~~ || . . .

Calculate $\dfrac{\text{number of sixes}}{\text{number of throws}}$ (as a decimal)

2. Toss a coin 100 times and record the number of heads.

Calculate $\dfrac{\text{number of heads}}{\text{number of tosses}}$

3. Put 7 red counters (or cubes or pegs) and 3 yellow ones in a bag or box. Without looking, take one out, note its colour and replace it. Do this 50 times.

Calculate $\dfrac{\text{number of red counters taken out}}{\text{total number of counters taken out}}$

4. Throw a die 50 times and record whether you get an even or an odd number.

Calculate $\dfrac{\text{number of evens obtained}}{\text{total number of throws}}$

5. Shuffle a pack of playing cards. Take out one card at random, note its suit, return it to the pack and shuffle again. Do this 40 times.

Calculate $\dfrac{\text{number of times a diamonds card is obtained}}{\text{total number of cards taken}}$

6. Drop 10 drawing pins onto a sheet of paper. Record the number of pins pointing upwards. Do this 10 times.

Calculate $\dfrac{\text{number of pins pointing upwards}}{\text{total number of pins dropped}}$

Suppose the following results were obtained for Experiment **2.** Six groups tossed a coin 100 times and recorded the number of heads.

Group	A	B	C	D	E	F	*Total*
No. of heads	55	52	45	58	47	40	297
No. of tosses	100	100	100	100	100	100	600
$\dfrac{\text{No. of heads}}{\text{No. of tosses}}$	0.55	0.52	0.45	0.58	0.47	0.40	0.495

The ratios $\frac{\text{no. of heads}}{\text{no. of tosses}}$ are shown on the graph.

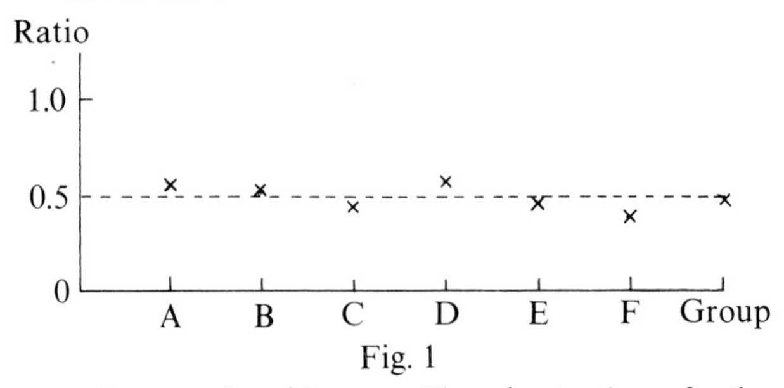

Fig. 1

Compare these results with yours. There is a tendency for the ratios to cluster around the value 0.5. Out of 600 tosses, 297 were heads. The ratio for the whole group is $\frac{297}{600} = \frac{99}{200} = 0.495$. This suggests that in the long run, the chance or probability of obtaining a head on one toss of a coin is 0.5.

Some results for Experiment **3** using 7 red and 3 yellow counters are given in the table below:

Group	A	B	C	D	E	*Total*
No. of red counters	32	37	30	35	39	173
Total no. of counters	50	50	50	50	50	250
Ratio $\frac{\text{no. of red}}{\text{total}}$	0.64	0.74	0.60	0.70	0.78	0.692

The ratios are shown on the graph.

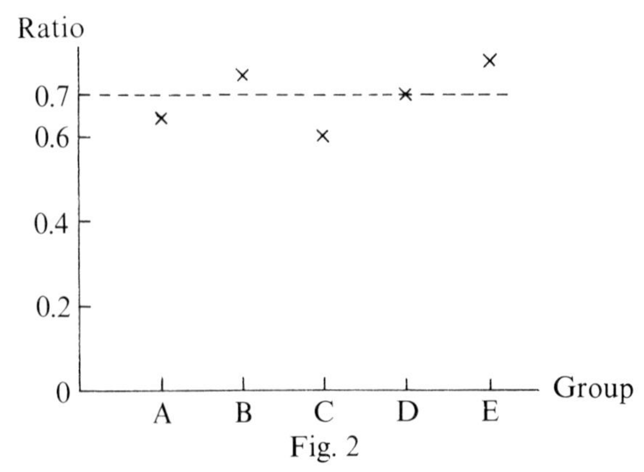

Fig. 2

Compare the results with yours. The results tend to cluster.

Out of 250 counters taken, 173 were red. This ratio $\frac{173}{250} = 0.692$. This is very close to the proportion of red counters in the bag ($\frac{7}{10} = 0.7$) and it suggests that, in the long run, the probability of picking a red counter is 0.7.

The work we have just done is based on experiments. Now let us see if we can work out the probability of an event.

Imagine that 4 red counters and 2 green ones are placed in a bag. If a counter is taken out, without looking, what is the probability that it is a red one? As there are twice as many red counters as green ones, it is reasonable to expect that if we repeat the draw 600 times we should get a red one about 400 times and a green one about 200 times. This suggests that the probability of drawing a red one is $\frac{400}{600} = \frac{4}{6} = \frac{2}{3}$. We write this as p(red) $= \frac{2}{3}$. Similarly, p(green) $= \frac{200}{600} = \frac{1}{3}$.

Now consider an ordinary die. When thrown, it can land in six different ways. Any of the numbers 1, 2, 3, 4, 5 or 6 can be on the top face and each is equally likely. In 600 throws of the die, we should expect to get the number 2 about 100 times. This suggests the probability of throwing 2 is $\frac{100}{600} = \frac{1}{6}$. We write p(2) $= \frac{1}{6}$. We should expect not to get 2 about 500 times and so p(not 2) $= \frac{500}{600} = \frac{5}{6}$. Notice that we assume each of the six possibilities is equally likely, that is, the die is not biassed. We speak of the set of possible outcomes and write it as $\{1, 2, 3, 4, 5, 6\}$. Now consider the probability of throwing less than 3. We say that 1 and 2 are the favourable outcomes. In 600 throws we should expect 1 about 100 times and 2 about 100 times and so a favourable outcome about 200 times. p(less than 3) $= \frac{200}{600} = \frac{2}{6} = \frac{1}{3}$.

Notice that $\frac{2}{6}$ is $\frac{\text{number of ways of getting less than 3}}{\text{number of possible ways the die can fall}}$.

In general,

$$\text{probability of an event} = \frac{\text{number of favourable outcomes}}{\text{total number of possible outcomes}}.$$

Exercise 25

1. State the set of possible outcomes for:

(i) throwing a die (ii) tossing a coin

(iii) writing a single digit number
(iv) picking a suit from a pack of playing cards.

2. Write in symbols: (for example p(red) $= \frac{2}{3}$)
(i) The probability of throwing 2 with a die is $\frac{1}{6}$.
(ii) The probability of obtaining a head when tossing a coin is $\frac{1}{2}$.
(iii) The probability of picking from the alphabet a letter that is a vowel is $\frac{5}{26}$.
(iv) The probability of taking an ace from a pack of cards is $\frac{4}{52} = \frac{1}{13}$.

3. If the set of possible outcomes is $\{1, 2, 3, 4, 5, 6, 7, 8, 9\}$, find:
(i) p(even) (ii) p(odd) (iii) p(not 6) (iv) p(less than 4).
What have you assumed?

4. Imagine that the faces of a die have the set of numbers $\{1, 2, 3, 4, 5, 7\}$.
Find: (i) p(odd) (ii) p(even) (iii) p(6) (iv) p(not 6).

5. A card taken at random from a pack is either a court card (king, queen or jack) or not a court card. What is the probability of taking a court card? What is the probability of not taking a court card? Explain why we cannot regard the set {a court card, not a court card} as a set of equally likely outcomes.

6. The whole numbers from 1 to 100 inclusive can be classified as less than 10, more than 10 or equal to 10. Explain why this is not a set of equally likely outcomes.

7. Discuss whether or not the set {win, lose, draw} for a football team playing at home is a set of equally likely outcomes.

IMPOSSIBLE EVENTS

Remember that

$$\text{p(event)} = \frac{\text{number of favourable outcomes}}{\text{total number of outcomes}}$$

If an event is impossible, there are no ways it can happen and so the number of favourable outcomes is zero. Hence the probability of the event is 0.

In one throw of an ordinary die, having the numbers 1 to 6, p(8) = 0.

When a counter is drawn from a box containing only red and green counters, p(blue) = 0.

CERTAIN EVENTS

If an event is certain to happen, then the number of favourable outcomes is equal to the total number of possible outcomes and so the probability is 1.

If a card is drawn from a pack containing only hearts and diamonds, p(red card) = 1.

When tossing a die, p(< 7) = 1, since it is certain that the number thrown is less than 7.

RANGE OF PROBABILITIES

The number line below illustrates the probabilities of some events using a die.

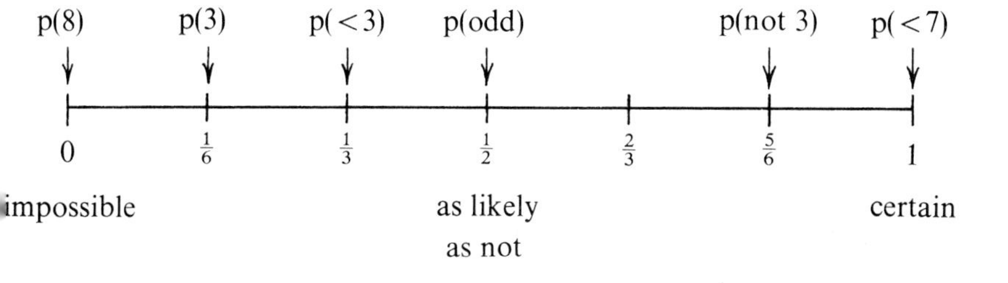

Fig. 3

This illustrates that the probability of an event is between 0 and 1 or is equal to 0 or 1. In symbols $0 \leqslant p \leqslant 1$.

EXAMPLES

1. A letter is picked at random from {a, e, i, o, u}. P(vowel) $= \frac{5}{5} = 1$ and p(consonant) $= \frac{0}{5} = 0$.

2. A class of twenty-eight pupils has two sets of twins. If a name is picked at random what is the probability it is the name of a twin? P(twin) $= \frac{4}{28} = \frac{1}{7}$. Also p(not twin) $= \frac{24}{28} = \frac{6}{7}$. Notice that p(twin) + p(not twin) $= \frac{1}{7} + \frac{6}{7} = 1$.

3. On page 59, we discussed the probability of picking a red counter from a bag containing 4 red and 2 green. The set of possible

outcomes could be {red, green} but this is not very helpful because the two colours are not equally likely. For equally likely outcomes we can number the counters and use the set

$\{r_1, r_2, r_3, r_4, g_5, g_6\}$. Then
$p(r) = \frac{4}{6} = \frac{2}{3}$, $p(g) = \frac{2}{6} = \frac{1}{3}$, $p(r \text{ or } g) = \frac{6}{6} = 1$,
$p(\text{blue}) = \frac{0}{6} = 0$.

Exercise 26

1. A class of 30 pupils has 14 boys. When choosing a name at random, what is the probability of getting a girl's name?

2. State an event which is certain and another event which is impossible.

3. Write down the set of outcomes for tossing a coin once. Find:
(i) p(head) (ii) p(head or tail) (iii) p(2 tails).

4. If the set of equally likely outcomes is {2, 3, 4, . . ., 11, 12} find:
(i) p(even) (ii) p(prime) (iii) p(odd or even)
(iv) p(1) (v) $p(\not> 4)$ (vi) p(even or prime).

5. 200 tickets were sold for a raffle. If a man bought 4 of them, what was his probability of winning the prize?

6. A letter is chosen at random from the word PROBABILITY. State the set of possible outcomes. Find:
(i) p(vowel) (ii) p(consonant) (iii) p(A or B)

7. A pile of tickets contains 20 red ones, 15 white and 5 blue. One is taken at random. Find the probability that it is:
(i) red (ii) white (iii) blue
(iv) red or blue (v) not white (vi) yellow.

8. A card is taken at random from a set of 52 playing cards. Find the probability that it is:
(i) a red card (ii) a heart (iii) an ace
(iv) not a diamond (v) a picture card
(vi) a court card (jack, queen, king).

9. A box contains equal numbers of triangles, squares, rectangles, pentagons and hexagons. If one shape is taken out, find the probability that:
(i) it is a triangle (ii) it has more than 4 sides
(iii) it is not a square (iv) it has 3 or 4 sides.

10. The set of counting numbers can be arranged in three subsets A, B and C where A = {3, 6, 9, ...}, B = {1, 4, 7, ...} and C = {2, 5, 8, ...}. Any positive integer written down at random belongs to A, B or C. Let {a, b, c} be the set of equally likely outcomes where $a \in A$, $b \in B$ and $c \in C$. Find:

(i) p(a) (ii) p(a or c) (iii) p(not b) (iv) p(b and c).

The set of counting numbers is {1, 2, 3, 4, 5, ...}. Write down any two and find their product. It is odd or even. Are these two outcomes equally likely? Look at this table.

First number	*Second number*	*Product*
odd	odd	odd
odd	even	even
even	odd	even
even	even	even

The product is even three times as often as it is odd. Therefore p(even product) = $\frac{3}{4}$ and p(odd product) = $\frac{1}{4}$.

Exercise 27

1. For the product of counting numbers just discussed, find:

(i) p(not even) (ii) p(not odd)
(iii) p(odd or even) (iv) p(odd and even).

2. Three cards, identical except for colour, are red, white and blue. Two cards are picked up without looking. Write down the three possible outcomes. Assuming these are equally likely, find the probability of:

(i) red and white cards (ii) no blue card
(iii) two white cards (iv) one red and one other.

3. A man has a pair of black shoes and a pair of brown shoes. K_L denotes the black left shoe. The others are called K_R, N_L and N_R. He takes two of the shoes out of a cupboard without looking at them. Complete this set of equally likely outcomes:

$$\{(K_L K_R), (K_L N_L), (K_L N_R), \ldots\}$$

State the probability of:

(i) one of each colour (ii) a pair
(iii) two right shoes (iv) at least one left shoe

4. On page 59, we saw that in throwing a die p(2) $= \frac{1}{6}$ and p(not 2) $= \frac{5}{6}$. Notice that p(2) + p(not 2) = 1. This illustrates the general result:

$$\text{p(event occurring)} + \text{p(event not occurring)} = 1$$

This gives a way of finding the probability that an event does not happen:

$$\text{p(event not occurring)} = 1 - \text{p(event occurring)}$$

Use this to find for a die:

(i) p(not 4) (ii) p(not even) (iii) p(not prime).

5. A set of equally likely outcomes is $\{1, 2, 3, 4, \ldots, 8, 9\}$. Find:

(i) p(8) (ii) p(not 8) (iii) p(even)
(iv) p(not even) (v) p(prime) (vi) p(not more than 5).

6. A seaside town claims that, out of every 6 days, 3 are sunny, 2 are cloudy and dry and 1 is rainy. For a day chosen at random we can write p(sunny) $= \frac{3}{6} = \frac{1}{2}$. Find:

(i) p(cloudy and dry) (ii) p(rainy)
(iii) p(not rainy) (iv) p(sunny or cloudy and dry).

7. The owner of a snack bar estimated the probability of a customer buying tea, coffee or a soft-drink as follows: p(tea) $= \frac{5}{12}$, p(coffee) $= \frac{4}{12}$, p(soft drink) $= \frac{3}{12}$. Find:

(i) p(not tea) (ii) p(not a soft drink) (iii) p(tea or coffee).

On page 58, we discussed Experiment 3. In that experiment, a counter was drawn many times from a bag containing 7 red and 3 yellow counters. Suppose that we draw a counter n times and get a red one d times. Then $\frac{d}{n}$ is near to 0.7, the probability of a red. With a very large number of draws (or trials), we should expect $\frac{d}{n}$ to be very close to 0.7. It follows that d would be very close to $n \times 0.7$. $n \times 0.7$ is called the *expected number* of reds for n draws (or trials). For 500 trials, the expected number of reds is $500 \times 0.7 = 350$. We should expect to get *about* 350 reds. We might, for example, get, 341 or 357. We should *not* expect to get a number a long way off 350, such as 105 or 410.

In general, if we call a certain result a 'success', the expected number of successes = number of trials × probability of a success.

EXAMPLES:

1. In 1000 throws of a coin, the expected number of heads = $1000 \times 0.5 = 500$.
2. In 300 throws of a die, the expected number of sixes is $300 \times \frac{1}{6} = 50$.
3. In taking a card from a shuffled pack 100 times, the expected number of cards that are hearts is $100 \times \frac{1}{4} = 25$.
4. If the probability of a rainy day in August is $\frac{2}{7}$, then in a fortnight's holiday the expected number of rainy days is $14 \times \frac{2}{7} = 4$.

Exercise 28

1. A die is thrown 120 times. How many times would you expect:
   (i) 3 to come up (ii) an odd number (iii) 1 or 6 to come up?
2. For a certain biassed coin, p(head) = $\frac{2}{5}$ and p(tail) = $\frac{3}{5}$. In 100 tosses how many heads would you expect? How many tails?
3. If you took a card 60 times from a shuffled pack, how many times would you expect *not* to get a diamond?
4. Over the years an annual flower show has had fine weather 4 times out of 5. Over a ten year period, how many wet flower shows would you expect?
5. Records show that on average 1 in 20 workers at a factory are absent on Mondays. Express this as a probability. If the firm has 800 workers, how many are expected to be absent each Monday?
6. A car park contains 80 cars. How many would you expect to have number plates that end in 7, 8 or 9?
7. A hotel estimates that the probability of being full any day is $\frac{3}{5}$. How many days per year would the hotel expect to be full?
8. At a snack bar, the estimated probabilities for a customer asking for tea, coffee or a soft drink are p(tea) = $\frac{5}{12}$, p(coffee) = $\frac{4}{12}$, p(soft drink) = $\frac{3}{12}$. Estimate the likely number of teas, coffees and soft drinks required by 600 customers.
9. Eight per cent of the applicants to a certain police force are rejected because their eyesight is not good enough. Out of 125 applicants, how many would you expect to be rejected for this reason?

8 · USING LOGARITHMS 1

BASIC IDEAS

Exercise 29

1. Write down the values of $2^0, 2^1, 2^2, 2^3, 2^4, 2^5$.

2. Copy and complete this table where $y = 2^x$

x	0	1	2	3	4	5	6	7	...	11	12
y	1	2			16						4096

3. Use your table to express 32, 128 and 4096 as powers of 2.

4. Write as a single power of 2:
(i) $2^3 \times 2^5$ (ii) $2^5 \times 2^7$ (iii) $2^{12} \div 2^4$ (iv) $(2^5)^3$.

5. Look carefully at the following statements:
$32 \times 128 = 2^5 \times 2^7 = 2^{5+7} = 2^{12} = 4096$
Use this method and the table of Question **2** to find the value of:
(i) 4×64 (ii) 128×8 (iii) 16×16 (iv) 32×64.

6. Study the following:
$4096 \div 256 = 2^{12} \div 2^8 = 2^{12-8} = 2^4 = 16$
Use this method and the table of Question **2** to find the value of:
(i) $512 \div 32$ (ii) $2048 \div 16$
(iii) $1024 \div 64$ (iv) $4096 \div 32$.

Look again at the statement

$$32 \times 128 = 2^5 \times 2^7 = 2^{5+7} = 2^{12} = 4096$$

With the aid of the table of powers of 2, we have found the answer to 32×128 by adding the indices 5 and 7.

In Question **6** of Exercise **29**, division questions were worked out by subtracting one index from another.

These are the basic ideas to be used in this chapter.

Instead of using powers of 2, we use powers of 10. Here is an example:

$$1000 \times 100 = 10^3 \times 10^2 = 10^5 = 100\,000$$

Any number can be expressed as a power of 10.

For example, $5.36 = 10^{0.7292}$ and $21.4 = 10^{1.3304}$

Hence, $5.36 \times 21.4 = 10^{0.7292} \times 10^{1.3304} = 10^{2.0596}$

which is 114.8

(This is an approximate answer only. The decimal parts of the powers of 10 are not exact but are correct to 4 sig. fig.)

Another name for index is *logarithm*.

Here is part of a table of logarithms:

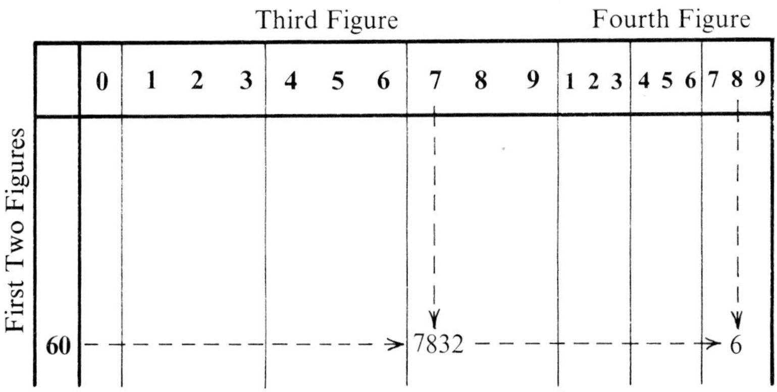

	Third Figure										Fourth Figure								
First Two Figures	0	1	2	3	4	5	6	7	8	9	1	2	3	4	5	6	7	8	9
60								7832										6	

The logarithm of 6.07 is 0.7832

that is $6.07 = 10^{0.7832}$

(Notice that decimal points are *not* given in the table.)

The logarithm of 6.078 is $0.7832 + 0.0006$, that is 0.7838,

hence $6.078 = 10^{0.7838}$.

Exercise 30

Use tables to write down the logarithms of:

1. 3.59 **2.** 3.595 **3.** 7.46 **4.** 7.467
5. 5.682 **6.** 3.137 **7.** 2.838 **8.** 1.7
9. 4 **10.** 8 **11.** 1.083 **12.** 8.903.

13. Use the above results to calculate 3.137×2.838.

14. Use tables and one of the above results to calculate 3.17×2.355.

NUMBERS GREATER THAN 10

The tables give the logarithms of numbers from 1 to 9.999. Logarithms of other numbers can be obtained by converting them to standard form.

EXAMPLES:

$$56.8 = 5.68 \times 10 \simeq 10^{0.7543} \times 10^1 = 10^{1.7543}$$

therefore, the logarithm of 56.8 is 1.7543

$$5680 = 5.68 \times 1000 \simeq 10^{0.7543} \times 10^3 = 10^{3.7543}$$

therefore, the logarithm of 5680 is 3.7543.

The logarithm of a number greater than 10 consists of two parts:

1. the whole number part which is called the *characteristic*,
2. the decimal part which is called the *mantissa*.

Exercise 31

Express the following in standard form and then state their logarithms:

1. 23.5, 235, 2350, 23 500.
2. 846, 84 600, 8 460 000, 84.6.
3. 10.8, 10 800, 108, 108 000.
4. 500, 5000, 50, 50 000.

You may have noticed that the characteristic can be found as follows:

Start with the decimal point in the standard form position and count the number of places you must move it in order to obtain the given number.

EXAMPLES: 63400 $\quad$ 6.3400 $\quad$ 4 places $\quad$ log 34000 = 4. . . .

852.6 $\quad$ 8.526 $\quad$ 2 places $\quad$ log 852.6 = 2. . . .

5. State the characteristics of the logarithms of the following numbers:

(i) 46.27 $\quad$ (ii) 63970 $\quad$ (iii) 8700
(iv) 927.95 $\quad$ (v) 30.61 $\quad$ (vi) 5.26.

Write down the logarithms of:

6. 7.92, 79.2, 792, 7920.

7. 5.23, 523, 52 300, 52.3.

8. 3.48, 3480, 34.8, 348 000.

9. 4, 4000, 400, 40 000.

10. 9.2, 920, 92, 920 000.

11. 8.352, 835.2, 835 200, 8 352 000.

12. 1.603, 16.03, 16 030, 160 300.

ANTILOGARITHMS

What number has 4.537 as its logarithm? That is, what number can be expressed as $10^{4.537}$?

We can use the logarithm tables in the reverse direction, but it is easier to use the special antilogarithm tables.

ANTILOGARITHMS

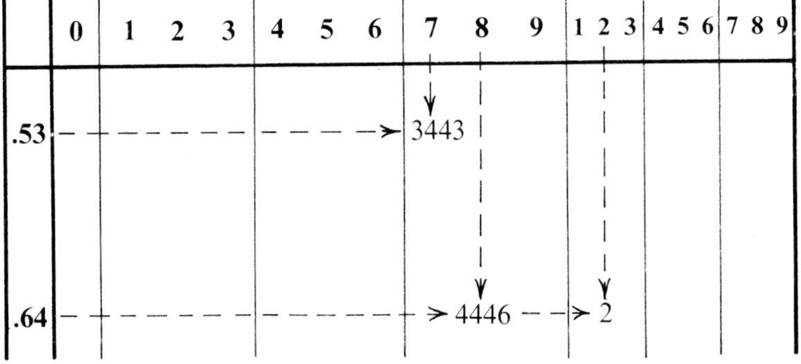

	0	1	2	3	4	5	6	7	8	9	1	2	3	4	5	6	7	8	9
.53								3443											
.64									4446			2							

Remember that the logarithm tables give only the decimal part and so we must look up only the decimal part in the antilogarithm table. This is shown on page 69.

For 0.537 we obtain 3.443 (The decimal point is not given in the table).

$\therefore 10^{0.537} = 3.443$

$10^{4.537} = 10^4 \times 10^{0.537} = 10\,000 \times 3.443 = 34\,430$

EXAMPLE: *Find the antilogarithm of 2.6482.*

$$10^{0.6482} = 4.448 \text{ from the antilog. tables}$$
$$10^{2.6482} = 10^2 \times 10^{0.6482} = 100 \times 4.448 = 444.8$$

Exercise 32

Find the antilogarithms of the following:

1. 0.8650, 3.8650, 1.8650, 2.8650.

2. 0.7540, 2.7540, 1.7540, 4.7540.

3. 0.4579, 1.4579. 3.4579, 5.4579.

4. 0.2577, 2.2577, 1.2577, 3.2577.

5. 0.9605, 3.9605, 1.9605, 4.9605.

6. 0.0472, 1.0472, 3.0472, 4.0472.

7. 0.0908, 1.0908, 4.0908, 2.0908.

8. 1.6150, 4.6150, 0.6150, 2.6150.

MULTIPLICATION AND DIVISION

EXAMPLE 1: $62.91 \times 4.83 \simeq 10^{1.7988} \times 10^{0.6839}$

$$\begin{array}{r} 1.7988 \\ +0.6839 \\ \hline 2.4827 \end{array}$$

$$= 10^{2.4827} = 10^2 \times 10^{0.4827}$$
$$\simeq 100 \times 3.039$$
$$= 303.9$$

(Rough check: $60 \times 5 = 300$)

EXAMPLE 2: $725.6 \div 38.3 \simeq 10^{2.8607} \div 10^{1.5832}$

$= 10^{1.2775} = 10^1 \times 10^{0.2775}$

$\simeq 10 \times 1.894$

$= 18.94$

(Rough check: $720 \div 40 = 18$)

$$\begin{array}{r} 2.8607 \\ -1.5832 \\ \hline 1.2775 \end{array}$$

The rough check shows you whether or not the decimal point is in the correct place.

Exercise 33

Use logarithm tables to calculate the following and carry out a rough check in each case:

1. 3.7×8.4 **2.** 53×7.6 **3.** 63.1×5.3
4. 6.8×78.3 **5.** $642 \div 28$ **6.** $935 \div 47$
7. $76.9 \div 3.4$ **8.** $83.2 \div 9.8$ **9.** 24.7×62.9
10. 3.59×29.6 **11.** 981×7.62 **12.** 74.3×2.76
13. $876 \div 92.5$ **14.** $2640 \div 17.9$ **15.** $1460 \div 28.4$
16. 43.91×2.75 **17.** 62.36×3.98 **18.** $39\,760 \div 4285$.

POWERS AND ROOTS

$(4.23)^3 \simeq (10^{0.6263})^3 = 10^{0.6263 \times 3}$

$= 10^{1.8789} \simeq 75.67$

It follows that $\sqrt[3]{75.67} \simeq 4.23$

To find $\sqrt[3]{75.67}$ by logarithms we must reverse the above working. Above, we multiplied 0.6263 by 3 to obtain 1.8789. We must now divide 1.8789 by 3 to obtain 0.6263

$$\sqrt[3]{75.67} \simeq 10^{1.8789 \div 3} = 10^{0.6263} \simeq 4.23$$

(Rough check: $4 \times 4 \times 4 = 64$)

Similarly for $\sqrt[4]{928}$ we divide the logarithm of 928 by 4

$$\sqrt[4]{928} \simeq 10^{2.9675 \div 4} = 10^{0.7419} \simeq 5.520$$

(Rough check: $5 \times 5 \times 5 \times 5 = 625$)

Exercise 34

Find the value of:

1. 14.5^3 **2.** 6.82^3 **3.** 2.37^4 **4.** 19.6^4

5. 1.88^5 **6.** $\sqrt[3]{37.6}$ **7.** $\sqrt[4]{982}$ **8.** $\sqrt[5]{72.7}$
9. $\sqrt[3]{5160}$ **10.** $\sqrt[4]{9.68}$ **11.** $\sqrt[5]{67\,200}$ **12.** $\sqrt[3]{52.8}$

RULES FOR WORKING WITH LOGARITHMS

1. *Multiplication.* $a \times b \times c \times \ldots$
   Add together the logarithms of $a, b, c, \ldots$.
2. *Division.* $a \div b$
   Subtract the logarithm of b from the logarithm of a.
3. *Powers.* a^p
   Multiply the logarithm of a by p.
4. *Roots* $\sqrt[q]{a}$
   Divide the logarithm of a by q.

ARRANGEMENT OF WORK IN COLUMNS.

It is laborious to write the numbers as powers of 10. An easier method is to set out the working in columns.

EXAMPLE 1:

62.91×4.83
$\simeq 303.9$
(Rough check: $60 \times 5 = 300$)

Number		*Logarithm*
62.91	→	1.7988
× 4.83	→	+ 0.6839
303.9	←	2.4827

EXAMPLE 2:

$(6.37)^3$
$\simeq 258.4$
(Rough check: $6^3 = 216$)

Number		*Logarithm*
6.37	→	0.8041
6.37^3	→	2.4123
258.4	←	

ACCURACY OF RESULTS

Tables give the logarithm of 6.37 to 4 significant figures as 0.8041. The logarithm of 6.37 to 6 sig. fig. is 0.804 139. Multiplying this by 3 we get 2.412 417 so that the 2.4123 in Example 2 above is not accurate in the last decimal place. Therefore the answer of 258.4 is not accurate in the last figure. In fact 6.37^3 is 258.474 853 which is nearer to 258.5 than 258.4.

Because the fourth figure is not reliable we usually give an answer to 3 sig. fig. when using 4 figure logarithms.

Exercise 35

Use logarithms to find the value of the following. Give each answer correct to 3 sig. fig.

1. 4.62×7.93
2. 5.87×8.26
3. 36.9×7.24
4. 84.6×4.68
5. $763 \div 92.1$
6. $684 \div 6.35$
7. $\sqrt[3]{765}$
8. $\sqrt[5]{9146}$
9. 3.62^3
10. 7.91^3
11. $469.7 \div 26.3$
12. $984.2 \div 19.6$
13. 30.4×12.6
14. 107.5×46.8
15. $\sqrt[4]{36.98}$
16. 3.429^4
17. 1.165^5
18. $\sqrt[3]{72\,940}$
19. $6.54 \times 7.98 \times 13.6$
20. $2.74 \times 12.9 \times 8.62$
21. $21.6 \times 1.34 \times 2.48$
22. $42.9 \times 7.42 \times 3.56$.

Exercise 36

Give the answers to 3 sig. fig.

1. Find the area of a rectangle 62.3 cm by 48.7 cm.

2. Find the area of a triangle of height 34.6 cm and base 23.4 cm.

3. Find the length of the edge of a cube of volume 325 cm^3.

4. Find the circumference of a circle of diameter 9.6 cm. (Use 3.142 for π)

5. Find the area of a circle of radius 17.4 cm.

6. Find the length of a rectangle of area 60.3 cm^2 and breadth 7.4 cm.

7. Find the mass of a cube of lead of edge 8.3 cm given that the mass of 1 cm^3 of lead is 11.4 g.

8. If the speed of sound in water is 1410 m/s, how long would it take for an echo to return from the bottom of the Mindanae Deep (10 900 m)?

9. Use logarithms to calculate 2^{20} and compare the answer with the exact value of 1 048 576.

10. Calculate the approximate value of 3^{30}, giving the answer in standard form.

REVISION PAPERS A

REVISION PAPER A1

1. $\xi = \{1, 2, 3, \ldots, 9\}$. $A = \{2, 4, 6, 8\}$, $B = \{3, 6, 9\}$.
Describe the sets A and B.
Show ξ, A and B in a Venn diagram and enter all the elements.
List: (i) A' (ii) $A \cup B$ (iii) $(A \cup B)'$ (iv) B' (v) $A' \cap B'$
Which two sets are the same?

2. (a) Write down the values of 5^3, 3^5, 6^2 and 2^6.

(b) Express in the form n^p where n is as small as possible:
81, 32, 625 and 10 000.

(c) Simplify: $a^4 \times a^5$, $b^8 \div b^2$, $c^4 \div c$ and $(d^6)^2$

3. (a) Calculate the fourth angle of a quadrilateral having angles of 72°, 103° and 112°.

(b) A pentagon has two angles of 90° and the other three are equal. Calculate the size of each.

(c) Calculate the interior angle of a regular 12-sided polygon.

4. (a) Reduce to their lowest terms: $\frac{15}{35}$, $\frac{6a^2}{4a}$ and $\frac{b^2 - 5b}{2b - 10}$.

(b) Solve: (i) $\frac{x}{3} + \frac{x}{5} = 8$ (ii) $\frac{2x}{3} - \frac{x}{4} = \frac{5}{2}$.

5. $A = \begin{pmatrix} 4 & 2 \\ 3 & -1 \end{pmatrix}$, $B = \begin{pmatrix} 6 & -5 \\ -4 & 3 \end{pmatrix}$ and $C = \begin{pmatrix} 6 \\ 7 \end{pmatrix}$

(a) Simplify where possible, A + B, B + C, BC, CA, AB and BA.

(b) Solve (i) X + A = B (ii) Y + B = 4A.

6. (a) Express as fractions in their lowest terms:
70%, 55%, 24% and $37\frac{1}{2}$%.

(b) Express as percentages: 0.3, 0.35, $\frac{3}{4}$ and $\frac{5}{8}$.

(c) Find the value of: (i) 45% of £3 (ii) $22\frac{1}{2}$% of £6.

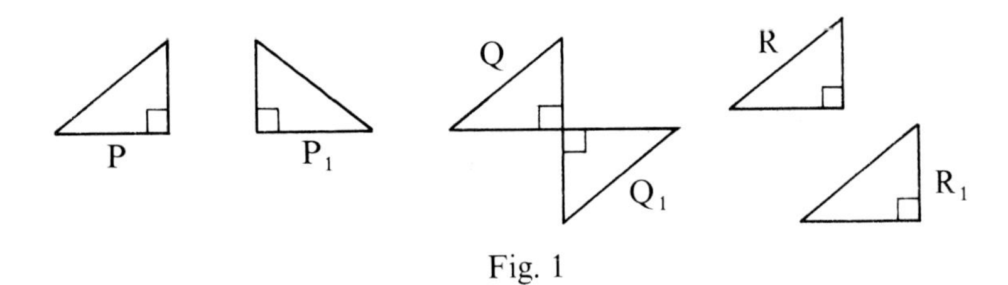

Fig. 1

7. (a) (i) In Fig. 1, what single movement (translation, reflection or rotation) takes the triangle from position P to position P_1?

(ii) What single movement takes it from Q to Q_1?

(iii) What single movement takes it from R to R_1?

(b) Fig. 2 shows the translations T_1 and T_2. Use squared paper to find the image of the point (4, 3) after applying the translation (i) $T_1 + T_2$ (ii) $T_1 - T_2$.

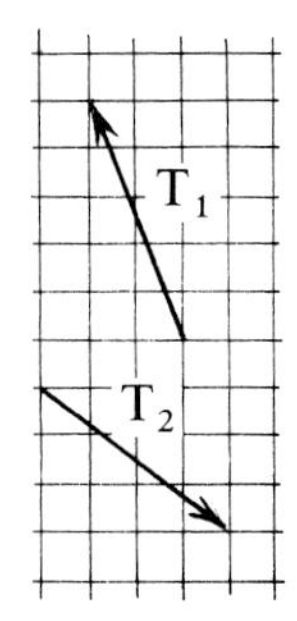

Fig. 2

8. Draw the graph of $y = x^2$ for values of x from -3 to $+3$. From your graph, find approximate values for 2.4^2, 1.3^2, $(-1.6)^2$, $\sqrt{5}$ and $\sqrt{2.5}$.

Using the same axes draw the graph of $y = 5 - x$ and state the coordinates of the points which are on both graphs.

REVISION PAPER A2

1. (a) Divide £8.50 between John and Andrew in the ratio 2:3.

(b) 120 small cubes have a mass of 37 g. Find, correct to the nearest gramme, the mass of 1000 such cubes.

2. (a) Write in standard form: 60 000, 5400, 0.0078 and 0.0003.

(b) Write without powers of 10: 4.23×10^4, 9×10^2, 6×10^{-3} and 5.4×10^{-2}.

(c) State in standard form the answers to:

(i) $(3 \times 10^4) \times (2 \times 10^2)$ (ii) $(5 \times 10^3) \times (7 \times 10^4)$

(iii) $(2.5 \times 10) \times (6 \times 10^2)$ (iv) $(7 \times 10^{-2}) \times (6 \times 10^{-3})$.

3. (a) Write down the squares of:

(i) $\frac{2}{3}$ (ii) $5a^3$ (iii) 0.2 (iv) $-1\frac{1}{4}$.

(b) State the square roots of:

(i) 0.09 (ii) $\frac{9}{25}$ (iii) $36b^{36}$ (iv) $7\frac{1}{9}$.

4. Use logarithms to calculate, correct to 3 sig. fig.:

(i) 82.4×3.97 (ii) $8240 \div 397$ (iii) 39.7^3.

5. Calculate the area of the field shown in Fig. 3 given that AC = 240 m, BE = 120 m and FD = 140 m.

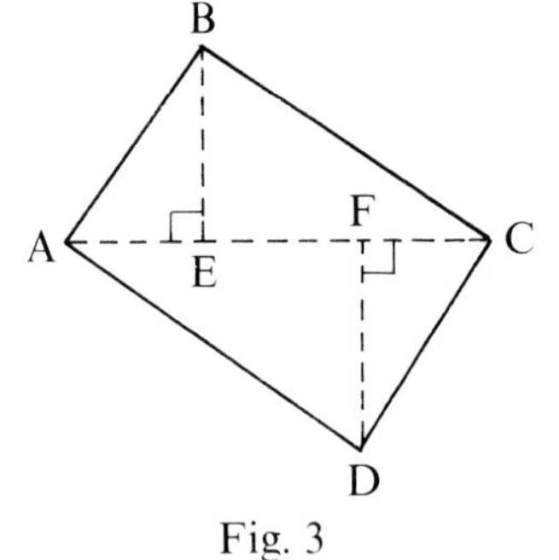

Fig. 3

6. (a) Simplify $5(2a - 3b) - 3(a - 2b)$.

(b) Solve $x(x - 5) = x(x + 2) + 3(x - 1)$.

(c) Solve $6(y + 2) - 2(y + 7) = 10$.

7. (a) A certain event is impossible. What number do we give to its probability? Give an example.

(b) A certain event must take place. What number do we give to its probability? Give an example.

(c) State the set of possible outcomes when a die is tossed and find:

(i) p(5) (ii) p(less than 5).

8. Make simple sketches to show the shapes of the graphs of:

(i) $y = x^2$ (ii) $y = \frac{6}{x}$ (iii) $y = x$

(iv) $y = 3$ (v) $y = 3 - x$.

REVISION PAPER A3

1. List: (i) A (ii) $A \cap B$ (iii) B' (iv) $(A \cup B)'$.

Which are true?

(i) $A \subset B$ (ii) $B \subset A$ (iii) $B' \subset A$
(iv) $A' \subset B$ (v) $\xi' = \phi$ (vi) $\xi \subset A$.

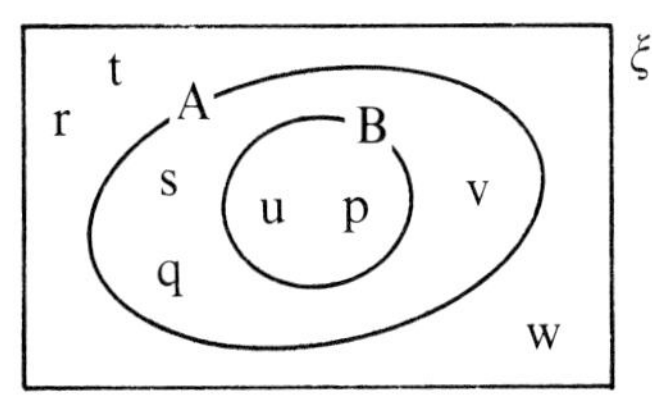

Fig. 4

2. (a) Simplify: $3a^3 \times 4a^4$, $10b^{10} \div 5b^5$; $(4c^3)^2$, $(d^{-4})^{-2}$.

(b) Solve the equations: $3^x = 81$, $2^y = \frac{1}{8}$, $5^z = 1$.

3. (a) State the logarithms of: 1000, 100 and 1.

(b) Use tables to calculate $(4.62)^4$ and $\sqrt[4]{4620}$.

4. (a) Simplify: $\frac{3}{5} + \frac{2}{3}$, $\frac{a}{b} + \frac{c}{d}$ and $\frac{4}{f^2} - \frac{2}{f}$.

(b) Solve $\frac{5x}{6} - \frac{2}{3} = \frac{3x}{4}$ and $\frac{y+4}{5} = \frac{y+5}{4}$

5. When a stone has fallen d metres its speed s m/s is given approximately by the formula $s = \frac{9}{2}\sqrt{d}$

(i) Calculate s when $d = 36$ and when $d = 100$.
(ii) Rearrange the formula so that d is the subject.
(iii) Calculate d when $s = 18$ and when $s = 90$.

6. (a) On squared paper copy and complete these diagrams to give the image of each object in the mirror M.

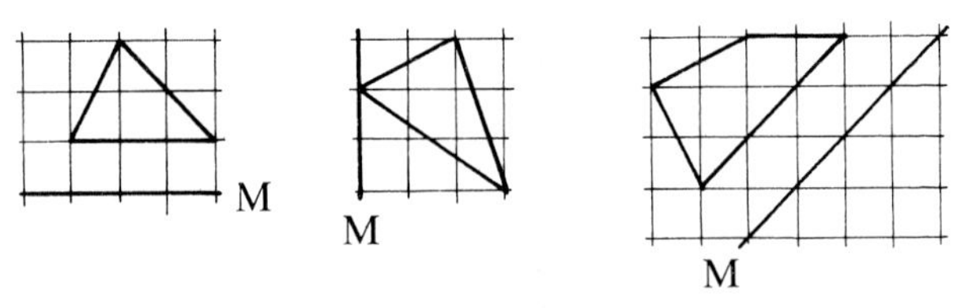

Fig. 5

(b) Draw triangle PQR on squared paper where P is the point (2, 1), Q is (5, 3) and R is (3, 4). In the line $y = x$ draw the reflection of triangle PQR and letter it $P_1Q_1R_1$. State the coordinates of P_1, Q_1, R_1. What do you notice about the coordinates of points and their reflections in the axis $y = x$?

7. Using $3\frac{1}{7}$ for π, calculate the circumference and area of a circle of diameter 42 cm.

8. (a) Calculate the hypotenuse of a right-angled triangle in which the other sides are 7 cm and 24 cm.

(b) Calculate the third side of a right-angled triangle which has a hypotenuse of 10 cm and a side of 5 cm.

REVISION PAPER A4

1. (a) An isosceles triangle has an obtuse angle. What can you say about each of the other two angles?

(b) Calculate a, b and c. Give reasons for your answers.

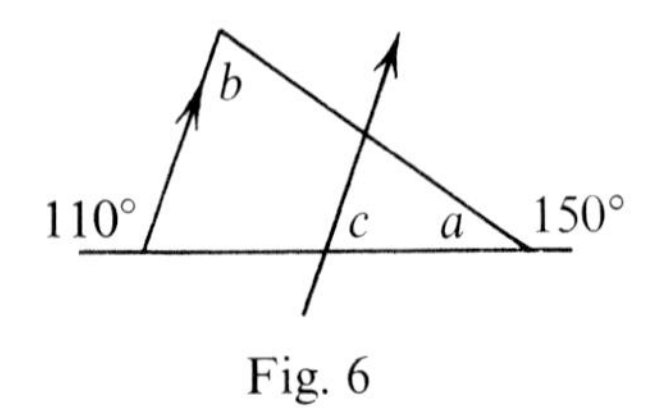

Fig. 6

2. (a) Express as decimals: 23%, 7%, $18\frac{1}{2}$% and $4\frac{1}{2}$%.
(b) Express as percentages: $\frac{2}{5}$, $\frac{3}{20}$, 0.9 and 0.035.

3. A box contains 12 red pens, 18 black pens and 20 blue pens. The pens are identical except for colour. One is taken out at random. Find the probability that it is:

(i) black (ii) not red (iii) blue or black (iv) neither red nor black.

4. (a) Simplify: $\frac{3}{5} \div \frac{9}{20}$, $\frac{a}{b^2} \div \frac{3}{b}$ and $\frac{c^2}{7} \div \frac{c}{21}$

(b) Solve: $\frac{3x - 2}{4} + \frac{x - 3}{6} = \frac{2x - 1}{3}$.

5. If a certain cafe owner serves n lunches, each lunch costs him C pence where $C = 60 + \frac{1500}{n}$.

(i) Explain this formula.
(ii) Calculate C when $n = 20$ and when $n = 100$.
(iii) Show that the formula can be arranged to give $n = \frac{1500}{C - 60}$.
(iv) Calculate n when $C = 90$ and when $C = 72$.
(v) If the owner charges 80p for each lunch, how many must he serve in order to make a profit?

6. (a) Which of the following are Pythagorean triads?

(i) 6, 8, 10 (ii) 12, 14, 20 (iii) 20, 21, 29
(iv) 16, 25, 36 (v) 12, 35, 37 (vi) 9, 40, 41.

(b) Using BD^2 (Fig. 7), calculate CD and the area of the quadrilateral.

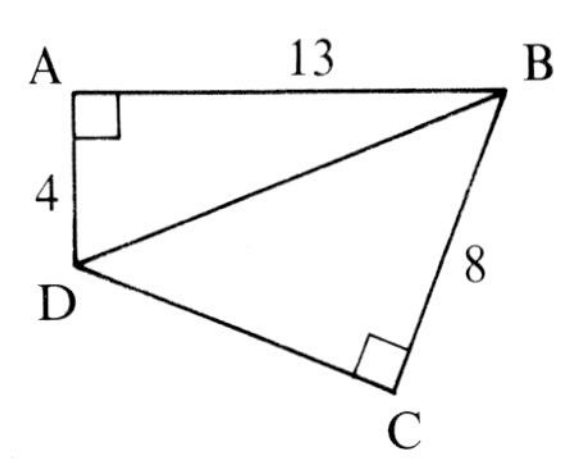

Fig. 7

7. Using $\pi = 3.14$, calculate the circumference and area of a circle of radius 5.4 cm. Give each answer to 2 sig. fig.

8. On squared paper draw the graph of $y = \frac{6}{x}$ for values of x from 1 to 6.

From your graph state the approximate answers to $6 \div 2.8$, $6 \div 1.7$ and $6 \div 3.3$.

Using the same axes draw the graph of $y = 5x - x^2$ for values of x from 0 to 5. State the coordinates of the point where y has its greatest value on this graph.

State the coordinates of the points where the two graphs intersect.

REVISION PAPER A5

1. (a) Write as 10^n: 1000, $\frac{1}{1000}$, 0.0001, 1 million.

(b) State as fractions: 3^{-2}, 5^{-1}, 2^{-3}, 4^{-2}.

(c) Simplify: $e^{-3} \times e^{-5}$, $(f^3)^{-4}$, $(g^{-2})^{-6}$, $h^{-5} \div h$.

2. Use logarithms to calculate:

(i) 1570×12.8 (ii) $15\,700 \div 128$ (iii) 1.57^3.

3. (a) I start with a number x, add 5 to it and multiply the answer by 3. The result is 4 times x. Form an equation for x and solve it.

(b) Solve the simultaneous equations $\begin{cases} 4x - 3y = 14 \\ 5x + 3y = 4 \end{cases}$

4. (a) Factorise: $5a + 15b$, $cd - d^2$ and $3f^2 - 6fg + 9fh$.

(b) Solve, where possible: $x^2 = 16$, $y^2 = -25$, $w^3 = 8$ and $z^3 = -27$.

5. Calculate the area of the pentagon by dividing it into rectangles and triangles.

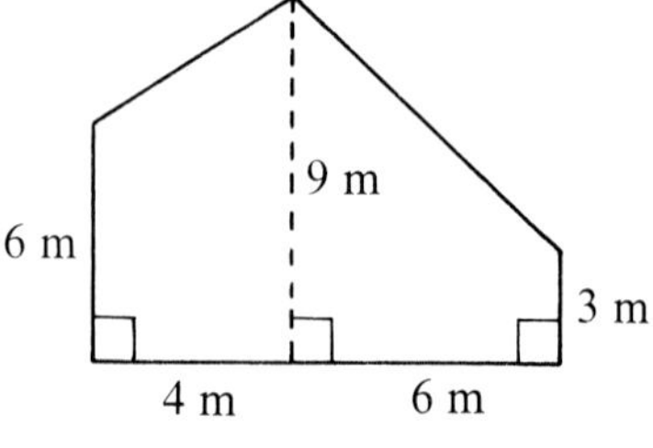

Fig. 8

6. (a) Cut out any triangle ABC from card (Fig. 9), place it on your paper and draw round it. Using the point of your compasses as a pivot, rotate the triangle through a convenient angle. Mark the new positions of A, B and C as A_1, B_1 and C_1.

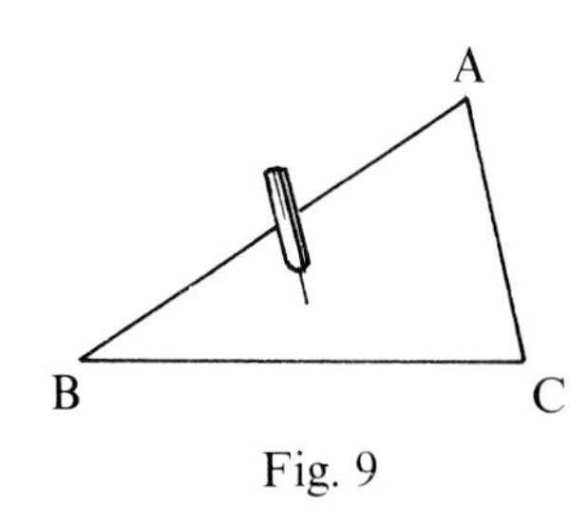

Fig. 9

Construct the mediators (perpendicular bisectors) of AA_1, BB_1 and CC_1. They should pass through the point of rotation.

(b) When rotated about a certain point in the plane, rod PQ takes up position P_1Q_1 (Fig. 10). Copy this diagram on squared paper and construct the point of rotation.

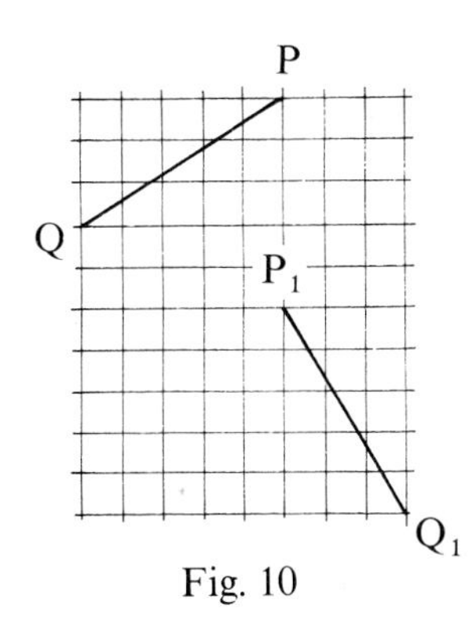

Fig. 10

7. Find the values of a, b, c, d, e, and f.

(i) $(4\quad 5)\begin{pmatrix}6\\a\end{pmatrix} = (34)$

(ii) $\begin{pmatrix}3 & 1 & b\\c & 2 & 1\end{pmatrix}\begin{pmatrix}1\\2\\3\end{pmatrix} = \begin{pmatrix}17\\12\end{pmatrix}$ (iii) $\begin{pmatrix}6 & 3\\f & 2\end{pmatrix}\begin{pmatrix}3 & e\\d & 5\end{pmatrix} = \begin{pmatrix}12 & 9\\11 & 5\end{pmatrix}$

8. The table shows the approximate amount of land per person in a certain country in different years. The amount of land is in units of 1000 square metres.

Year	1100	1300	1500	1600	1700
Amount of land	117	77	46	34	24
Year	1801	1851	1901	1951	2001
Amount of land	15	7.7	4.3	3.1	?

Represent the data graphically and comment on the information.

9 · MEAN AND MEDIAN

The weekly pocket money of five boys is 50p, 60p, 70p, 70p, and 100p; that of six girls is 40p, 50p, 50p, 60p, 70p and 90p.

The figure shows the two sets of money piled up in 10p coins. In the boys' pile is shown the individual amounts and what each boy would have if the money were shared equally among them. The girls' money has been arranged in the same way.

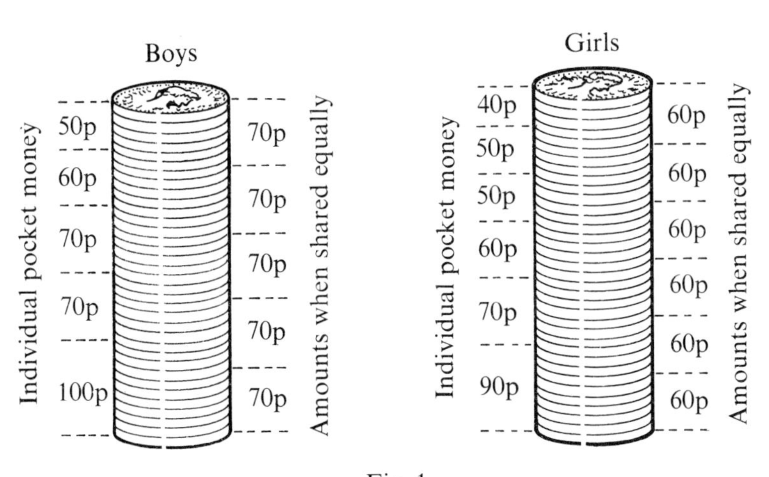

Fig. 1

By sharing the boys' money equally, we obtain the *mean* of the boys' pocket money.

$$\text{Mean of boys' pocket money} = \frac{\text{total of boys' money}}{\text{number of boys}}$$

$$= \frac{(50 + 60 + 70 + 70 + 100)}{5}\text{p}$$

$$= \frac{350\text{p}}{5} = 70\text{p}$$

$$\text{Mean of girls' pocket money} = \frac{\text{total of girls' money}}{\text{number of girls}}$$

$$= \frac{(40 + 50 + 50 + 60 + 70 + 90)\text{p}}{6}$$

$$= 60\text{p}$$

The means of 70p and 60p help us to compare the two sets of pocket money. The boys' mean is 10p higher than the girls'.

The term average is often used for the mean. In everyday usage, the word average sometimes has a vague meaning as in 'the average man'. For this reason we shall use mean rather than average. The mean of a set of values can be used to represent the set when comparing it with other sets.

Here are two other examples:

The ages of four children in a family are 5, 7, 9 and 11 years.

$$\text{The mean age} = \frac{\text{total of ages}}{4} = \frac{5 + 7 + 9 + 11}{4} = \frac{32}{4} = 8 \text{ years}$$

In London the hours of sunshine recorded for the days of a certain week were 2, 5, 7, 4, 0, 8 and 9.

$$\text{Mean} = \frac{\text{total hours of sunshine}}{\text{number of days}}$$

$$= \frac{2 + 5 + 7 + 4 + 0 + 8 + 9}{7} = \frac{35}{7} = 5 \text{ hours}$$

Exercise 37

1. Find the means of these sets of numbers:

(i) 6, 7, 8, 9, 10 (ii) 17, 18, 21, 25, 27, 30

(iii) 0.9, 1.2, 1.3, 1.5, 1.6, 1.6, 1.7

(iv) 26, 27, 27, 28, 29, 29, 30, 32.

2. Calculate the mean pocket money, correct to the nearest penny, of the eleven children considered in the text before this exercise.

3. The number of pages in a daily newspaper one week were 26, 32, 28, 30, 36 and 28. Calculate the mean number of pages per day.

4. In one week the attendances in two classes were:

3 W:	24,	24,	25,	25,	28,	27,	29,	28,	30,	30
3 S:	25,	25,	28,	28,	26,	26,	21,	21,	25,	25

Calculate the mean attendance for each class.

5. Twelve girls and six boys took a typing course. At one stage in the course, the number of words typed per minute were:

Girls:	12,	19,	13,	16,	21,	10,	13,	21,	14,	20,	13,	20
Boys	10,	12,	14,	16,	11,	15						

Calculate the mean score for the girls and for the boys. What is the mean score for the 18 pupils?

6. In a novel of 237 pages, the mean number of words per page is 230. Find the number of words in the novel to the nearest thousand.

7. Startown football team scored 27 goals in 9 matches and Topton scored 29 goals in 10 matches. Which team has the better mean (or goal average)? Each team scored 4 goals in its next match. Which team then had the better mean?

8. On a match box was printed 'Average contents 95'. How many matches would you expect to find in 6 such boxes?

9. The mean of four numbers is 37. What is the sum of the numbers? Invent four such numbers, having a mean of 37.

10. In June and July, one bowler took 36 wickets for 225 runs and another took 40 wickets for 256 runs. Calculate the mean number of runs per wicket for each bowler. Who seemed to be the better bowler? If each took his next wicket for 31 runs calculate the new means and comment on the results.

11. If 2 340 000 stones, each of 2.4 tonnes, were used to build a pyramid, find the mass of stone used. Assuming it took 20 years to build the pyramid, find the mean mass of stone placed in position each week. Use a working year of 50 weeks.

Fig. 2 on page 86 shows seven pupils placed in order of height. As a representative height for this set of pupils, we can take the height of the middle pupil, Jane. This is 154 cm. It is called the *median* height.

The median is the middle value when values are arranged in order of size.

Fig. 2

EXAMPLE 1: *Find the median or middle value of the scores 8, 9, 10, 12, 14, 15, 15, 16, 17.*

These are already in order of size. The median is 14.

EXAMPLE 2: *Arrange these eight values in order and find the median: 5, 8, 10, 5, 2, 12, 6, 9.*

The values in order are: 2, 5, 5, 6, 8, 9, 10, 12.

The two middle values are 6 and 8. The median is taken as the average of these.

$$\text{Median} = \frac{6 + 8}{2} = 7$$

Exercise 38

1. State the median of each of these sets of scores:

(i) 1, 4, 5, 7, 7 (ii) 16, 17, 19, 20, 25, 25, 26

(iii) 10, 15, 20, 20, 30, 30, 35, 40

(iv) 0.2, 0.2, 0.3, 0.5, 0.7, 0.7.

2. Find the median mass of seven pupils whose masses, in kilogrammes, are:

41, 49, 40, 43, 46, 52, 45

3. In a cricket match, the members of one team scored

37, 101, 12, 19, 28, 25, 0, 8, 6, 6, 0.

Arrange the scores in order and state the median score. Also calculate the mean score. Which do you think is the better representative number?

4. Find the median value in each set:

(i) £12, £10, £8, £3, £12, £1, £15, £11, £13
(ii) 7.6 km, 2.3 km, 1.7 km, 9.2 km, 4.1 km, 8.6 km, 7.4 km, 6.5 km, 2.9 km
(iii) 30 g, 56 g, 46 g, 40 g, 22 g, 38 g, 27 g, 53 g, 57 g, 49 g, 35 g, 48 g, 36 g
(iv) 6 h, 7 h, 4 h, 12 h, 15 h, 6 h, 3 h, 9 h.

5. The scores of eight people playing darts were: 25, 10, 42, 160, 27, 6, 12, 38. Find the median score. What is the mean score? Which do you think is the better representative score?

6. The pocket money, in pence, of nine children was:

60, 90, 40, 50, 120, 25, 120, 150, 65

Find the mean and median pocket money. Which do you think is the better representative for this group?

7. In a certain term the number of days absence for the pupils in class 3Z were:

0, 0, 0, 0, 0, 0, 0, 0,
4, 6, 6, 7, 7, 7, 8, 8, 10, 10, 10, 11, 14, 27, 28

Find the median and mean number of days' absence. Which do you think is the better presentation of the data?

The ages of five children are 9, 4, 10, 6, 11 years. The mean age is $(9 + 4 + 10 + 6 + 11) \div 5 = 40 \div 5 = 8$ years.

In 10 years time their ages will be 19, 14, 20, 16, 21 years. The mean age will then be $(19 + 14 + 20 + 16 + 21) \div 5 = 90 \div 5 = 18$ years. Notice that the mean age also will have increased by 10 years.

Three years ago their ages were 6, 1, 7, 3, 8 years. The mean age was $25 \div 5 = 5$ years, which is 3 years less than the present mean age.

When the same amount is added to (or subtracted from) each of a set of measurements, the mean is increased (or decreased) by the same amount.

Exercise 39

1. The ages of the members of a family are 40, 37, 16, 14, 12 and 7 years. Find the mean age:

(i) now (ii) 5 years ago (iii) in 7 years' time.

2. A firm employed eight salesmen. Each received £3000 salary per year plus commission. Last year the commission paid was

£1140, £830, £1020, £1000, £750, £900, £1090, £1270.

Calculate the mean commission and then state the mean amount received. Compare this with the median amount received.

3. Seven children have weekly pocket money of 40p, 36p, 60p, 55p, 50p, 48p, 40p. At Christmas each gets an increase of 20p. Find the original mean and the new mean.

4. On twenty channel crossings, the number of passengers carried by a ferry were:

387, 410, 479, 428, 443, 467, 486, 492, 535, 494,
532, 458, 523, 475, 420, 461, 394, 440, 487, 409

The ferry also carried a crew of 19. Calculate the mean number of people on these crossings.

5. One year the mean annual wage of a firm's employees was £2900. As a result of a productivity agreement, wages rose by 12% the following year. Calculate the new mean annual wage.

6. Find the mean and median values of this set of numbers:

28, 25, 30, 43, 20, 45, 23, 21, 23, 22

What are the values of the mean and median:

(i) when 12 is added to each number
(ii) when 20 is subtracted from each number
(iii) when each number is doubled?

The above exercise suggests that when values are close together there is a quicker method of calculating the mean.

EXAMPLE 1: *The ages at which five men retired were: 68, 65, 63, 62, 67 years. Calculate the mean.*

The smallest age is 62 years. Subtracting 62 from each age we get the numbers 6, 3, 1, 0, 5. The mean of these numbers is

$$\frac{6 + 3 + 1 + 0 + 5}{5} = \frac{15}{5} = 3$$

The mean retiring age is $3 + 62 = 65$ years.

We have really found the mean of the men's ages 62 years ago as 3 years and then found the mean now by adding on the 62 years. We say that the number 62 is a *working origin*. By reducing the size of the original numbers we have an easier calculation.

EXAMPLE 2: *The numbers of records and cassettes sold at a shop in each of six months were: 202, 196, 208, 190, 204, 212. Calculate the mean using a working origin of 200.*

Subtracting 200 from each of the given numbers we obtain

$$2, -4, 8, -10, 4, 12.$$

Mean of reduced numbers is

$$\frac{2-4+8-10+4+12}{6} = \frac{26-14}{6} = \frac{12}{6} = 2$$

The mean number of records is $2 + 200 = 202$.

In Example 2, two of the reduced numbers are negative because the original numbers were smaller than the working origin. The two number lines below show the original numbers and the reduced numbers.

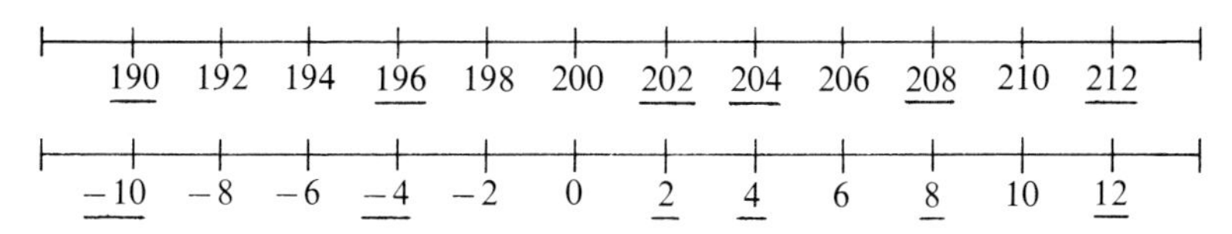

When choosing a working origin there are two alternatives:

1. We can choose a number which is a rough estimate of the mean. For example, for the ages of 30 pupils in a class we might look at the class list and choose a working origin of 14 years.

2. We can choose a number which enables differences to be found easily. For example, for 1035, 1027, 996, 1003, . . . we would choose 1000.

Exercise 40

1. For each part of this question, draw two parallel number lines. On one line show the original numbers; on the other show the corresponding reduced numbers, using the given working origin.

   Find the mean of the reduced numbers and hence of the original numbers.

   (i) 78, 71, 74, 73, 79. Use a working origin of 70.
   (ii) 267, 258, 261, 255, 260, 262, 257. Use 260.
   (iii) 12.2, 12.7, 10.9, 11.7, 12.0, 10.5, 13.4, 10.2. Use 12.0.

2. Over a period of 10 years the number of trees planted annually in a district was:

   1710, 1980, 2120, 2170, 1830, 2000, 2130, 1940, 2250, 1670.

   Choose your own working origin and calculate the mean, What is the median number?

3. The heights, in centimetres, of 11 boys and 7 girls are:

Boys:	155, 150, 158, 143, 170, 148, 157, 141, 147, 151, 163
Girls:	136, 140, 145, 134, 162, 142, 156

   Compare the two sets by finding their means and medians.

4. Ten samples of soil were analysed. The percentages of moisture content were:

   6.8, 9.2, 7.3, 6.5, 8.1, 5.9, 10.7, 8.7, 6.8, 9.0.

   What were the mean and median percentage moisture contents?

5. A doctor recorded the ages in years of 12 widows and 8 widowers who came to him for treatment. They were:

Widows:	75, 60, 69, 72, 58, 79, 64, 86, 65, 82, 70, 84
Widowers:	64, 75, 62, 80, 65, 70, 76, 60

   Compare the two sets by finding the mean and median ages of each set. Comment on your results.

6. In a certain class the marks obtained in Mathematics by 14 boys and 15 girls were:

Boys:	38,	54,	65,	72,	58,	76,	41,	33,
Boys:	56,	89,	46,	35,	56,	65		
Girls:	49,	56,	68,	38,	52,	75,	79,	55,
Girls:	48,	64,	28,	85,	43,	58,	72	

   For each set of marks find the mean and median and comment on your results.

7. The lives, in hours, of sixteen electric light bulbs were:
   237, 294, 268, 320, 285, 249, 243, 275, 242, 229, 206, 217, 258, 300, 270, 283.
   Does this sample match the maker's claim that the mean life of a bulb is 250 hours?

8. At a school, the times, in seconds, for the 100 metres races were:

Boys:	12.0,	12.4,	12.7,	12.7,	13.2,	13.8
Girls:	13.5,	13.8,	14.4,	14.8,	15.2,	15.9

Calculate the boys' mean time and the girls' mean time. Comment on the results. What is the speed in kilometres per hour of each winner?

EXAMPLE: *The mean pocket money of 15 girls is 45p and of 12 boys is 63p. Find the mean pocket money of the 27 children.*

For the girls,

$$\text{mean} = \frac{\text{total}}{15} = 45\text{p and so total} = 15 \times 45\text{p} = 675\text{p}$$

For the boys,

$$\text{mean} = \frac{\text{total}}{12} = 63\text{p and so total} = 12 \times 63\text{p} = 756\text{p}$$

For the whole group,

$$\text{mean} = \frac{675\text{p} + 756\text{p}}{27} = \frac{1431\text{p}}{27} = 53\text{p}$$

Exercise 41

1. The mean height of 14 boys is 156 cm and the mean height of 16 girls is 159 cm. Calculate the mean height of the 30 children.

2. At a pensioners' club the mean age of 42 women is 72 years and the mean age of 28 men is 67 years. What is the mean age of the whole group?

3. The mean mass of 28 pupils in a class was 47.5 kg. Twins joined the class. Their masses were 45.8 kg and 46.2 kg. What was the new mean mass of the class?

4. The mean length of 15 metal rods is 2.4 cm. The length of another rod is 4 cm. What is the mean length of the 16 rods?

5. The mean daily amounts of sunshine one June was 7 hours. How many hours of sunshine were there in that month?

6. At Exton-on-sea, the mean daily hours of sunshine for four months of 1977 were:

June 7.2, July 5.6, August 6.4, September 3.8.

Find the total number of hours of sunshine for each month. Calculate the mean number of hours per day over the whole period of four months.

7. The mean mass of 20 girls is 40 kg and that of 10 boys is 46 kg. Calculate:

(i) the combined mass of the 20 girls
(ii) the combined mass of the 10 boys
(iii) the mean mass of 30 boys and girls.

8. The mean height of 10 men is x cm. The mean height of 20 women is y cm. Find an expression for the mean height of the 30 people.

10 · HISTOGRAM AND MODE

A class collected data on the number of people in their families and obtained this frequency table.

Number in family (x)	3	4	5	6	7
Frequency (f)	4	12	8	4	2

Fig. 1 shows a histogram representing this information. Each frequency is represented by the area of a rectangle. The vertical scale on the left is an area scale showing that a square centimetre represents 4 people.

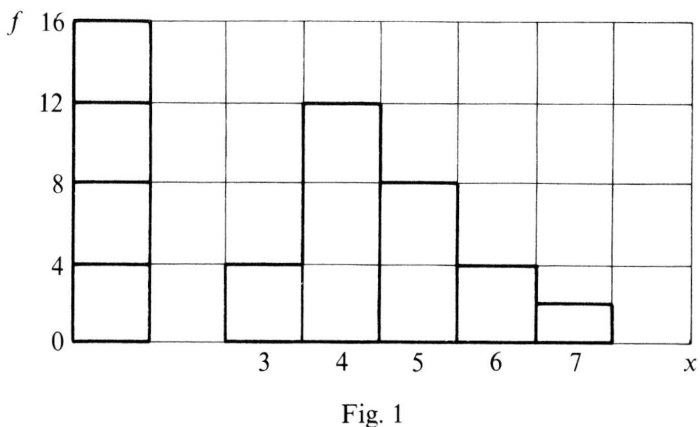

Fig. 1

The most common family size is 4 as it has the highest frequency. This is called the *mode.*

The next table shows the distribution of ages of employees in a certain firm.

Age in years (x)	16–20	21–30	31–40	41–50	51–60	61–64
Frequency (f)	35	100	140	180	90	32

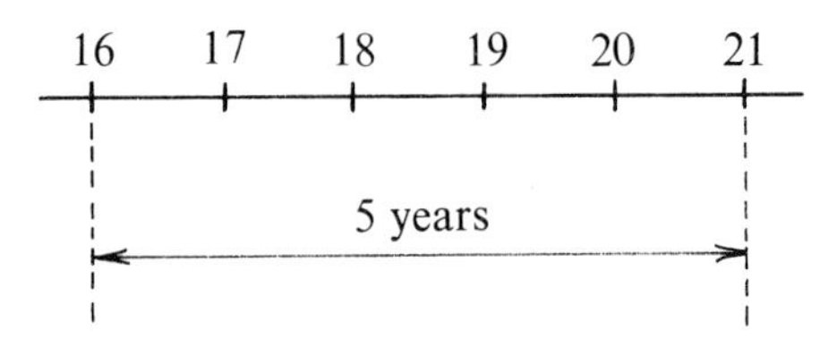

Fig. 2

As Fig. 2 shows, the interval 16–20 years means from 16 up to, but not including, 21 years. Thus the interval spans 5 years. Each of the next four intervals spans 10 years. The last interval of 61–64 years spans 4 years.

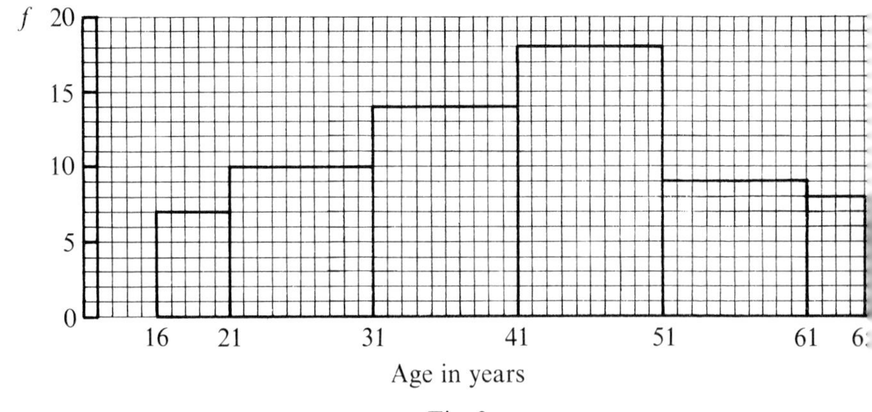

Fig. 3

On the area scale of the histogram in Fig. 3, one small square represents one person. The height for each rectangle in the histogram was found by dividing the area needed by the base. For example, the area of the first rectangle is to be 35 and its base is 5. Therefore its height is $35 \div 5 = 7$. For the last rectangle the area is to be 32, the base is 4 and so the height is $32 \div 4 = 8$.

For this distribution we have a *modal class* instead of a mode. It is the class 41–50 years as this has the highest frequency.

Exercise 42

For each histogram, use an area scale.

1. The frequency table shows the goals scored in matches one Saturday.

Number of goals (x)	0	1	2	3	4	5	6
Frequency (f)	26	39	16	5	1	0	1

Draw a histogram and comment on the results. Shade the area which represents at least 2 goals scored. What is the mode?

2. From the data in Question 1, calculate the total number of goals scored.

3. The number of letters received in the homes of members of a class in a certain week are given in this table:

Number of letters (x)	1	2	3	4	5	6	7	8
Frequency (f)	1	9	5	3	3	4	3	2

Represent the data by a histogram. What is the mode?
What proportion of homes received 1 or 2 letters?

4. The table shows the ages of buses in a certain company.

Age in years (x)	0–1	1–3	3–5	5–9
Frequency (f)	12	36	20	16

The class interval 1–3 is for buses aged 1 year or more but less than 3 years. Draw a histogram. What is the modal class?
Shade the area showing buses that are three years old or more. What percentage of the whole area is this?

5. The grades obtained by 50 pupils in a science examination were as follows:

5 4 2 3 4 1 5 4 3 5 4 6 2 6 2 4 4
4 1 3 4 6 5 3 5 4 4 5 3 4 3 6 4 1
3 5 2 5 5 2 3 4 6 4 2 5 3 4 4 2

Prepare a frequency table and draw a histogram. State the mode and comment on the results. Shade the area representing those with grades 3 and 4. What percentage is this?

6. Fig. 4 shows the histogram of the frequency distribution of money saved in one year by members of a small building society. On the area scale one square represents 20 people. By finding the areas of the rectangles draw up a frequency table. What is the modal class? Comment on the distribution.

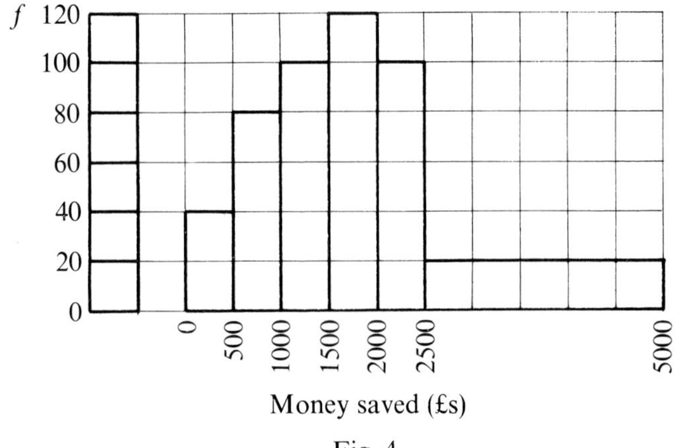

Fig. 4

7. In a certain factory the frequency of the number of working days' absence one month is given by the table:

Number of days' absence (x)	0	1	2	3	4	5	6	7
Frequency (f)	10	5	8	30	12	6	8	1

Draw a histogram. What is the mode? Comment on the distribution.

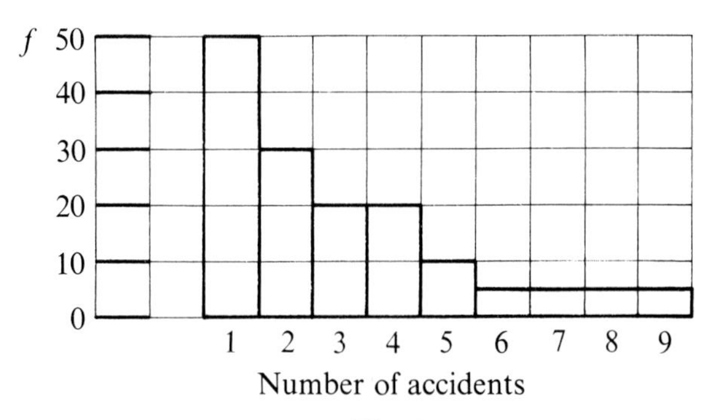

Fig. 5

8. The histogram in Fig. 5 represents the frequency distribution of accidents at a chemical plant one year. Thus 50 people had one accident, 30 people had 2 accidents and so on. Use the areas of the rectangles to prepare a frequency table. What is the mode? Calculate the total number of accidents. Comment on the distribution.

9. The table shows the masses of the parcels posted one day by a certain firm.

Mass in kilograms (x)	Less than 1	1–2	2–4	4–8	8–16
Frequency (f)	20	36	12	8	4

The class interval 4–8 contains masses of 4 kg or more but less than 8 kg. What is the modal class? Draw a histogram.

The variables in Questions **1** and **3** of Exercise **42**—number of goals and number of letters—can take only certain values. Such variables are called *discrete* variables.

Some other variables can have any value within certain limits. They are called *continuous* variables. Examples are the masses of parcels (Question **9** of Exercise **42**) and heights of people. If a height is given as 150 cm to the nearest centimetre, then it is between 149.5 cm and 150.5 cm. If a journey is stated to take $2\frac{1}{2}$ hours to the nearest half-hour, then it takes between $2\frac{1}{4}$ and $2\frac{3}{4}$ hours.

Exercise 43

1. Which of the following are variables:

(i) The number of centimetres in a metre.
(ii) The number of pupils absent from school.
(iii) The length of daylight.
(iv) The minimum voting age?

2. State whether each variable is discrete or continuous:

(i) The number of peas in a pod.
(ii) The mass of sugar in a basin.
(iii) The amount of water used each day in a house.
(iv) The length of a garden.
(v) The number of houses in a road.
(vi) The time to swim a length of a swimming bath.
(vii) Shoe size.

3. Name one variable that is discrete and another that is continuous. Explain how a discrete variable differs from a continuous variable.

4. Give the limits between which each of the following measurements lie:

(i) 48 m to the nearest metre
(ii) 16° to the nearest degree
(iii) 73.5 kg to the nearest $\frac{1}{2}$ kg.
(iv) 12.4 s to the nearest 0.2 second
(v) 9.6 cm to the nearest mm
(vi) 730 km to the nearest 10 km.

5. Give your height to the nearest centimetre and your mass to the nearest kilogramme. State the limits between which each measurement lies.

The table shows the heights, measured to the nearest centimetre, of 60 entrants to the police force.

Height in cm (x)	171–173	174–176	177–179
Frequency (f)	3	17	20
Height in cm (x)	180–182	183–185	186–188
Frequency (f)	10	8	2

If a man's height is 171 cm to the nearest centimetre, then it lies between 170.5 and 171.5 cm. Hence the three heights in the 171–173 cm interval are actually between 170.5 and 173.5 cm. These are called the *actual class limits.* As the difference of the actual class limits is 3 we say that the class interval is 3. For the histogram in Fig. 7 the actual class limits are used.

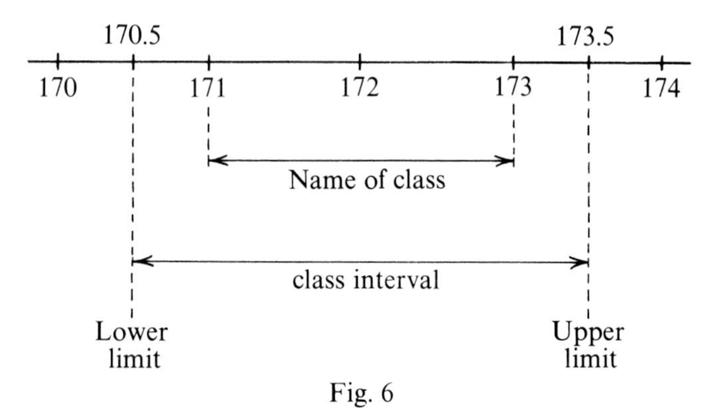

Fig. 6

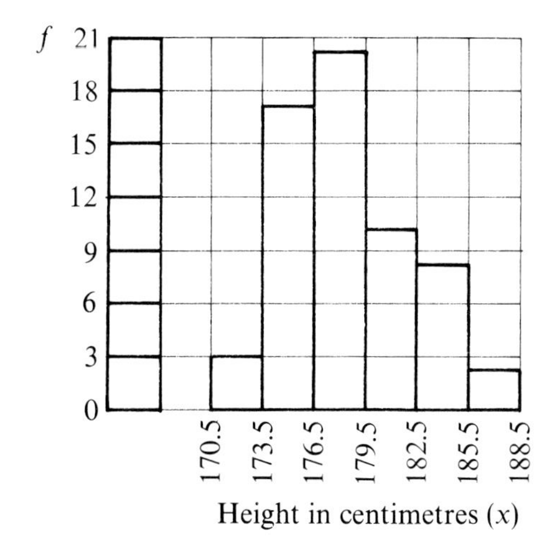

Fig. 7

Height is a continuous variable. Discrete variables can be grouped in the same way. For example, amounts of pocket money can be grouped into such classes as 50–59p and 60–69p.

Exercise 44

1. The table shows the amount of sleep, to the nearest hour, taken during one week by 100 men.

No. of hours of sleep (x)	28–33	34–39	40–45
Frequency (f)	5	6	14
No. of hours of sleep (x)	46–51	52–57	58–63
Frequency (f)	42	24	9

What is the class interval? Represent the data by a histogram using actual class limits. Name the modal class. Comment on the distribution. How much sleep do you have in a week?

2. The table on page 100 shows some examination results. What is the class interval?

Marks (x)	10–16	17–23	24–30	31–37
Frequency (f)	8	11	12	24
Marks (x)	38–44	45–51	52–58	
Frequency (f)	18	10	3	

State the modal class and draw a histogram. Comment on the distribution.

3. In one year a dentist treated 800 children. He recorded the number of teeth he removed from each child and obtained this table:

No. of teeth removed (x)	0	1	2	3	4	5
Frequency (f)	275	300	125	50	30	20

Draw a histogram and comment on the data.
Calculate the total number of teeth removed and so find the mean number of teeth removed per child.

4. 90 children were asked how far they lived from their school and this table was constructed:

Distance in kilometres (x)	Less than 1	1–3	3–5	5–10
Frequency (f)	8	32	40	10

The class interval 1–3 is for distances which are more than 1 km but less than 3 km. Draw a histogram, name the modal class and comment on the data.

5. The times taken by 80 children to swim one length of a swimming pool were recorded and this table constructed:

Time in seconds (x)	25–34	35–44	45–54
Frequency (f)	11	22	27
Time in seconds (x)	55–64	65–74	75–84
Frequency (f)	12	5	3

State the class interval and the modal class. Draw a histogram and comment on the data.

11 · THE TANGENT RATIO

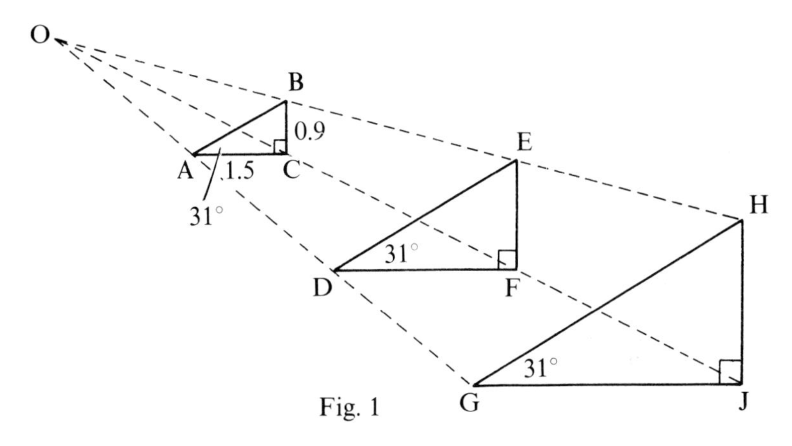

Fig. 1

In Fig. 1, $\triangle DEF$ is an enlargement of $\triangle ABC$ using centre O and scale factor 2. $\therefore EF = 2 \times 0.9 = 1.8$ and $DF = 2 \times 1.5 = 3.0$.

$\triangle GHJ$ is an enlargement of $\triangle ABC$ using a scale factor of 3. Hence $HJ = 3 \times 0.9 = 2.7$ and $GJ = 3 \times 1.5 = 4.5$.

$$\frac{BC}{AC} = \frac{0.9}{1.5} = \frac{3}{5} = 0.6, \qquad \frac{EF}{DF} = \frac{1.8}{3.0} = 0.6, \qquad \frac{HJ}{GJ} = \frac{2.7}{4.5} = 0.6$$

Suppose $\triangle KLM$ is an enlargement of $\triangle ABC$ using a scale factor of 7. Then $LM = 7 \times BC = 7 \times 0.9$ and $KM = 7 \times AC = 7 \times 1.5$.

Hence $$\frac{LM}{KM} = \frac{7 \times 0.9}{7 \times 1.5} = \frac{0.9}{1.5} = 0.6.$$

For all triangles which are enlargements of $\triangle ABC$, height ÷ base will be 0.6

In any right-angled triangle having an angle of 31°, height ÷ base = 0.6.

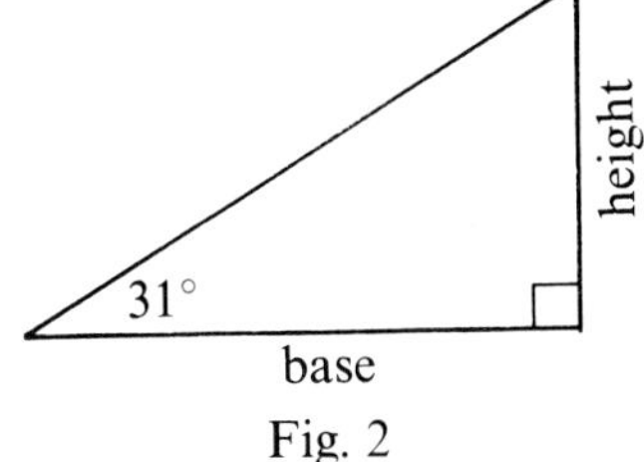

Fig. 2

Exercise 45

Draw the triangles in this exercise on graph paper.

1. Draw three triangles having angles of 31° and 90° and bases of 5 cm, 6 cm and 8 cm. For each triangle measure the height and divide it by the base. Each answer should be 0.6, correct to 1 sig. fig.

2. Draw three right-angled triangles having angles of 40° and bases of convenient lengths such as 5 cm, 8 cm and 10 cm. For each triangle measure the height and divide it by the base. Give each answer to 2 sig. fig.

3. Draw right-angled triangles having bases of 10 cm and angles of 25°, 30°, 35°, 50°, 55° and 60°.

You can draw all the triangles on one base if you wish as in Fig. 3. In each case measure the height to the nearest millimetre and state the value of height ÷ base.

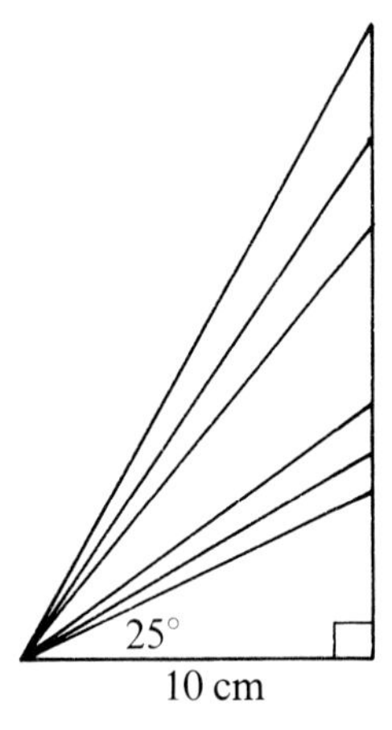

Fig. 3

What should the result be for a triangle with an angle of 45°?

Copy and complete the following table:

Angle	20°	25°	30°	35°	40°
Height ÷ base	0.36	0.47			
Angle	45°	50°	55°	60°	65°
Height ÷ base					2.14

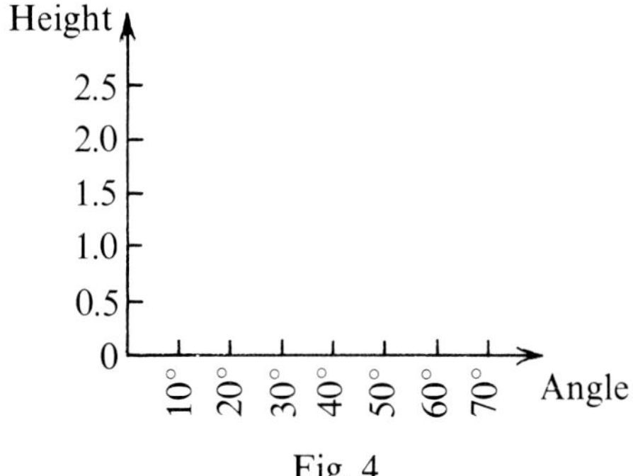

Fig. 4

4. Copy Fig. 4 on graph paper using 2 cm for 10° on the 'angle' axis and for 0.5 on the other axis. Plot points to represent the results of Question **3** and draw a smooth curve through the points. Using your graph copy and complete the table below.

Angle	32°	43°	58°	62°		
Height ÷ base					0.78	1.07

5. For all right-angled triangles with angles of 25°, height ÷ base $\simeq$ 0.47. For a triangle with a base of 3 m, $\dfrac{\text{height}}{3\text{ m}} \simeq 0.47$,

$\therefore$ height $\simeq 0.47 \times 3\text{ m} = 1.41$ m.

Calculate the heights of triangles having bases of 5 cm, 40 cm, 6 m and 20 m.

6. Use your table of Question **3** to find the heights in the following cases:

(i) base 20 cm, base angle 55°.
(ii) base 8 cm, base angle 35°.
(iii) base 4 cm, base angle 60°.

7. Use your graph of Question **4** to find the heights in the following cases:

(i) base 10 cm, base angle 37°.
(ii) base 30 cm, base angle 53°.

8. A right-angled triangle has a base of 20 cm and a height of 9.8 cm. Copy and complete:

Height ÷ base = ... ÷ ... = ... (as a decimal)
Base angle = ... (from your graph)

Use the same method to find the base angle if:
(i) the base is 10 cm and the height is 12 cm.
(ii) the base is 20 cm and the height is 14 cm.
(iii) the base is 5 m and the height is 7.4 m.

9. Calculate, to the nearest metre, the heights of the following objects given the angles of elevation at distances of 10 m from them:

(i) a lamp post, 35°
(ii) a tree, 63°
(iii) a window of a house, 46°.

We have seen that for each angle there is a number which can be used for calculations involving right-angled triangles. This number is called the *tangent* of the angle.

Thus the tangent of 25° is 0.47 approximately.

We write this as tan 25° $\simeq$ 0.47.

If θ is an angle in a right-angled triangle,

$$\tan\theta = \frac{\text{opposite side}}{\text{adjacent side}}$$

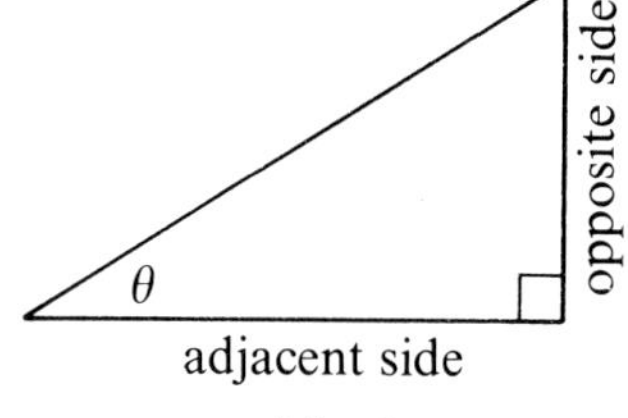

Fig. 5

It is possible to calculate the tangent of an angle to any number of decimal places. Fig. 6 shows part of a table which gives tangents to four decimal places. The tangent of 49° is 1.1504.

NATURAL TANGENTS

Degrees	0′	6′	12′	18′	24′	30′	36′	42′	48′	54′	Mean Differences in minutes				
		0.1°	0.2°	0.3°	0.4°	0.5°	0.6°	0.7°	0.8°	0.9°	**1′**	**2′**	**3′**	**4′**	**5′**
49	1.1504			1626							14				

Fig. 6

A degree is divided into 60 equal parts called *minutes*. ($1° = 60'$). The tangent of $49°18'$ (or $49.3°$) is 1.1626. (Notice that only the decimal part appears in the 18' column.) The tangent of $49°20'$ is $1.1626 + 0.0014 = 1.1640$.

Use tables to check your results in Questions **3** and **4** of Exercise **45**.

We shall now use tables to calculate unknown lengths in right-angled triangles.

EXAMPLE 1:

$$\frac{a}{16} = \tan 23° \simeq 0.4245$$

$$a \simeq 16 \times 0.4245 = 6.7920$$

BC is 6.79 cm to 3 sig. fig.

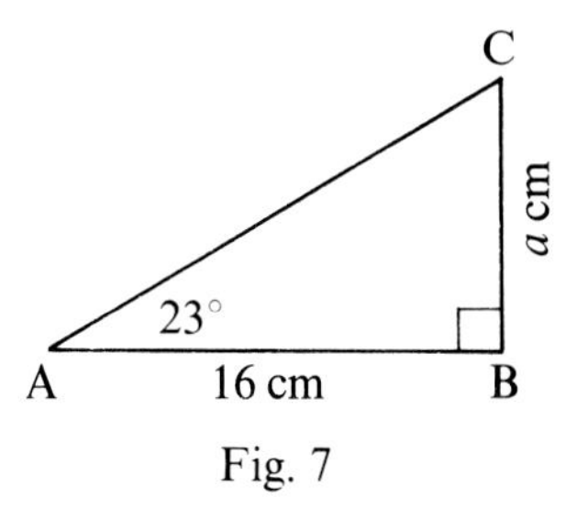

Fig. 7

EXAMPLE 2:

$$\frac{d}{7} = \tan 54° \simeq 1.3764$$

$$d \simeq 7 \times 1.3764 = 9.6348$$

EF is 9.63 cm to 3 sig. fig.

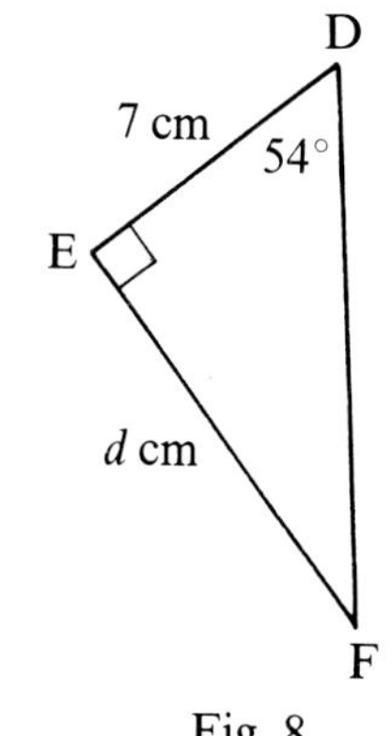

Fig. 8

Exercise 46

1. From tables write down the values of:

(i) $\tan 26.2°$ (ii) $\tan 43.4°$ (iii) $\tan 5.6°$
(iv) $\tan 56°$ (v) $\tan 75°42'$ (vi) $\tan 39°52'$

2. Use tables to find the tangents of 7°, 12°, 30°, 36°, 60° and 70°.
Is $\tan 60° = 2 \times \tan 30°$? Is $\tan 36° = 3 \times \tan 12°$?
Is $\tan 70° = 10 \times \tan 7°$?

Questions **3** to **11**. Calculate the sides labelled with small letters, giving your answers to 3 sig. fig.

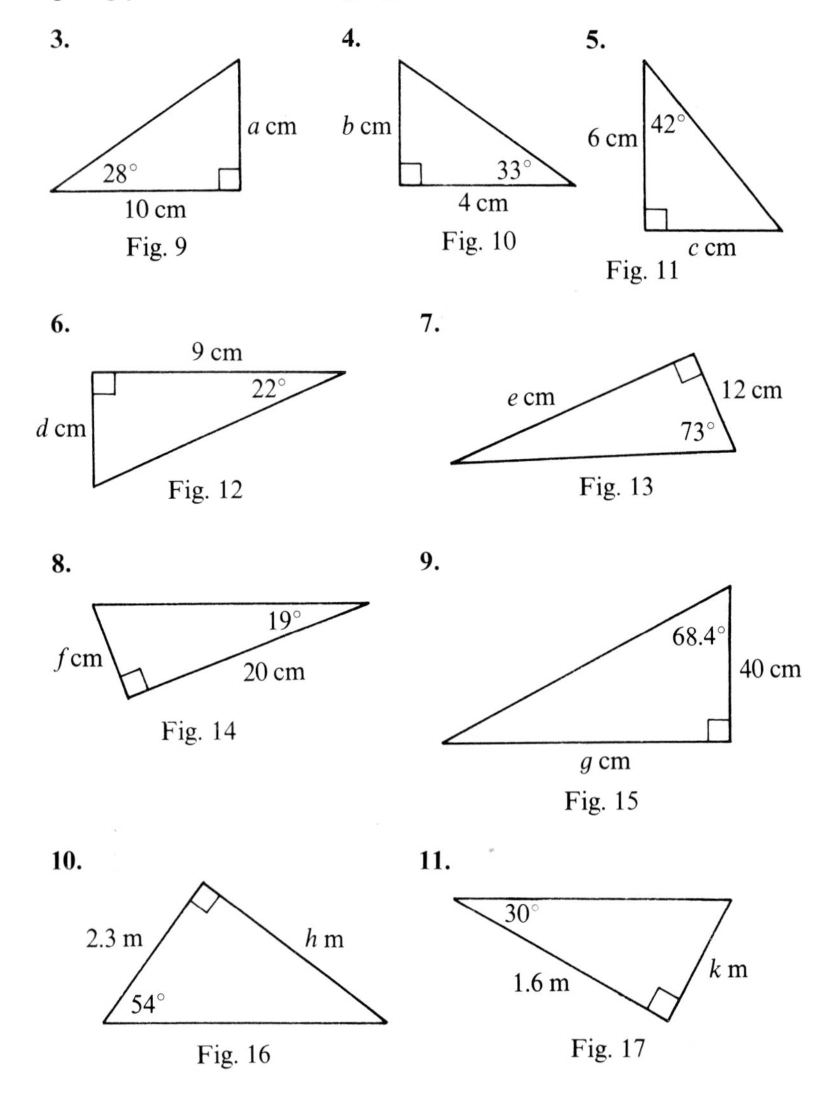

Fig. 9
Fig. 10
Fig. 11
Fig. 12
Fig. 13
Fig. 14
Fig. 15
Fig. 16
Fig. 17

Questions **12** to **14**. Calculate the other angle of the triangle and then the side labelled with a small letter.

12.

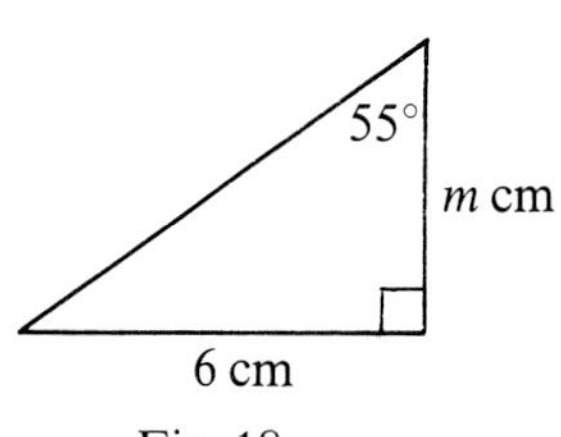

Fig. 18

13.

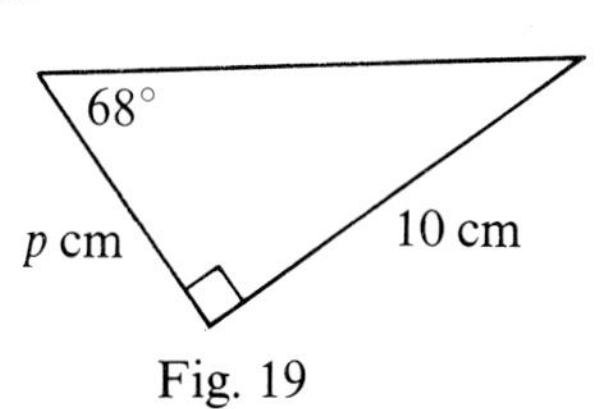

Fig. 19

14.

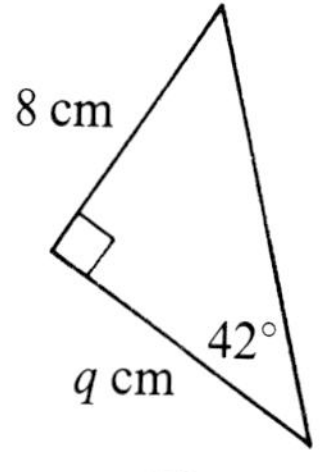

Fig. 20

15. Calculate GH in Fig. 21.

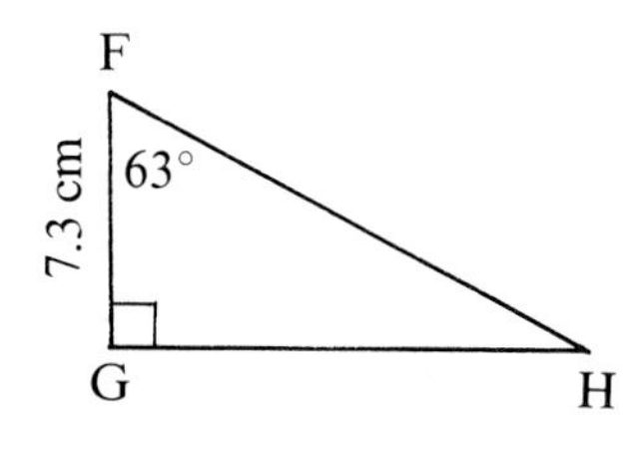

Fig. 21

16. In Fig. 22, calculate
(i) BC (ii) BD

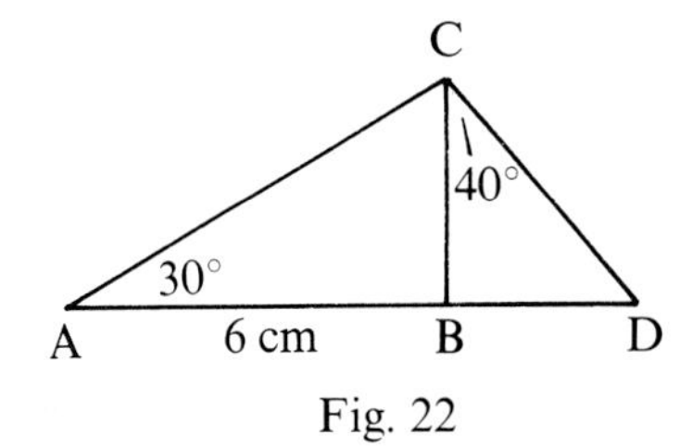

Fig. 22

17. Copy and complete the table below. Obtain the tangents of each angle from tables of tangents and round off each tangent to 2 decimal places. Draw the graph of $y = \tan x^\circ$ using 2 cm for 10° on the x-axis and 2 cm for 1 unit on the y axis.

Angle (x°)	0	10	20	30	40	50
Tangent (y)	0		0.36			
Angle (x°)	55	60	65	70	75	80
Tangent (y)						5.67

Is the graph a straight line? What connection has this with the statements in Question **2**?

PROBLEMS

Exercise 47

For each question, draw a neat diagram.

1. The foot of a ladder is placed 4.2 m from the wall of a house. The ladder makes an angle of 68° with the ground. How far up the wall does it reach?

2. At a distance of 20 m from a church tower, the angle of elevation of the top is 59°. Calculate the height of the tower.

3. In Fig. 23, the cliff is 50 m high and from its top the angle of depression of a small boat is 34°. State the size of angle θ. Calculate the distance of the boat from the foot of the cliff.

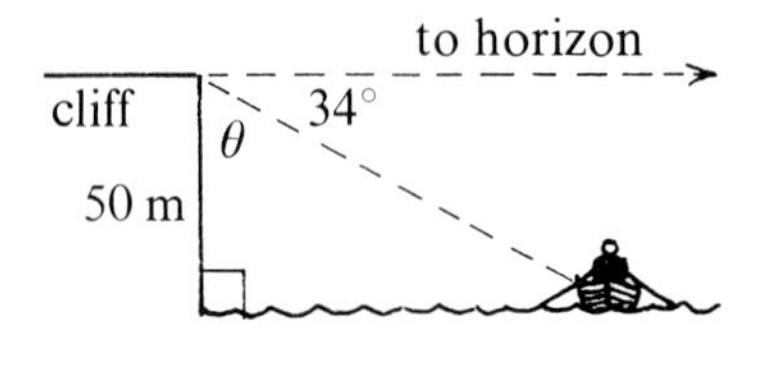

Fig. 23

4. Fig. 24 shows an isosceles triangle. Calculate the vertical height and the area of the triangle.

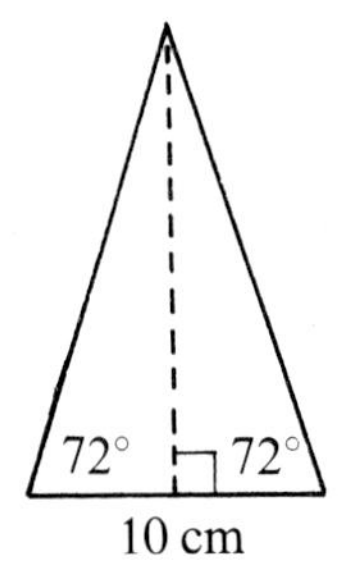

Fig. 24

5. The length of the base of an isosceles triangle is 16 cm and the angle opposite the base is 50°. Calculate the vertical height of the triangle.

6. Fig. 25 shows how a mast is to be supported by three wires. Each wire is to be attached to the mast at a height of 20 m from the base and is to make an angle of 60° with the ground.

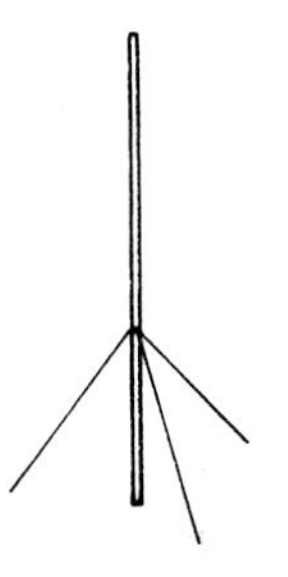

Fig. 25

How far should the end of each wire be placed from the foot of mast?

7. Standing on the top of a mountain 800 m high, a walker sees another peak which is at a horizontal distance of 2500 m from him and at an angle of elevation of 12°. Find the height of the other peak, correct to the nearest 100 m.

8. At a distance of 70 m from the base of a tower I measure the angle of elevation of the top as 32°. If my eye is 1.50 metres above ground level, calculate the height of the tower.

9. Standing at point A on a canal bank I see a post P on the opposite bank and find that AP is at 54° to the bank. I walk along the bank to B where the angle between PB and the bank is also 54°.

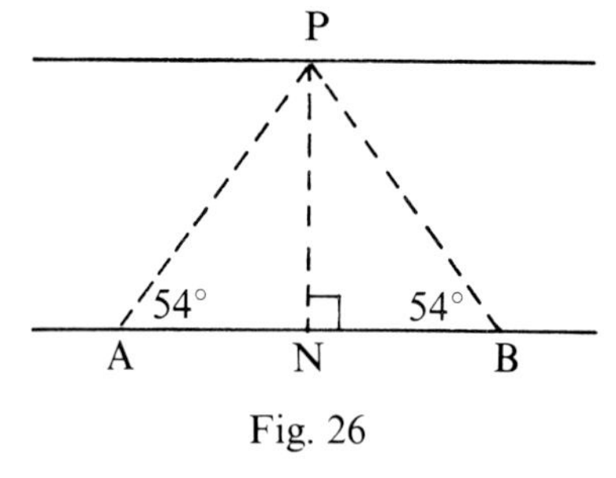

Fig. 26

If AB is 22 metres, calculate the width of the canal.

10. Fig. 27 shows a tower of height 20 m and a river AB. From T the angles of depression of A and B are 62° and 38°. Calculate: (i) AC (ii) BC and (iii) the width of the river.

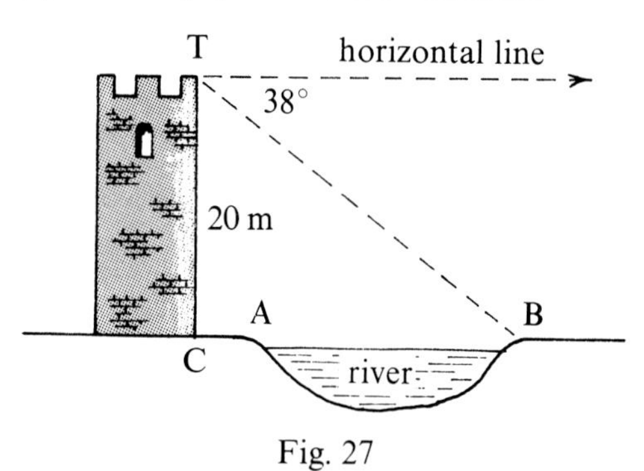

Fig. 27

11. A helicopter hovers vertically above a boat which is 600 m from shore. The angle of elevation of the helicopter from the shore is 36°. Calculate the height of the helicopter.

CALCULATING AN ANGLE OF A RIGHT-ANGLED TRIANGLE

EXAMPLE

$\tan\theta = \frac{3}{7} \simeq 0.4286$

$\theta \simeq 23.2°$ or $23°12'$

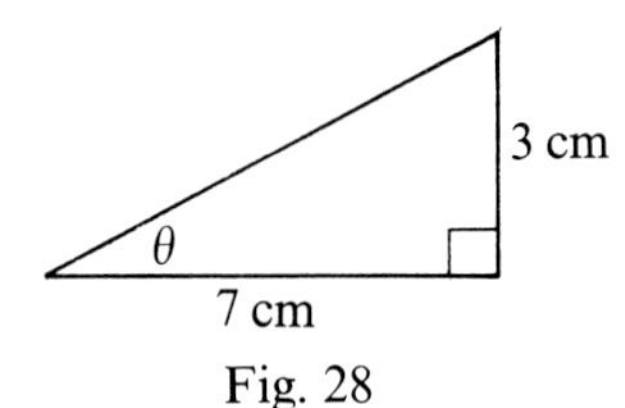

Fig. 28

Exercise 48

1. Use the tables to find angles having tangents of:
(i) 0.3057 (ii) 0.4813 (iii) 2.2889
(iv) 0.6548 (v) 3.1748 (vi) 1.7064.

Questions **2** to **9**: For each figure, state the value of tan θ to 4 decimal places and use your table of tangents to find the size of angle θ.

2.

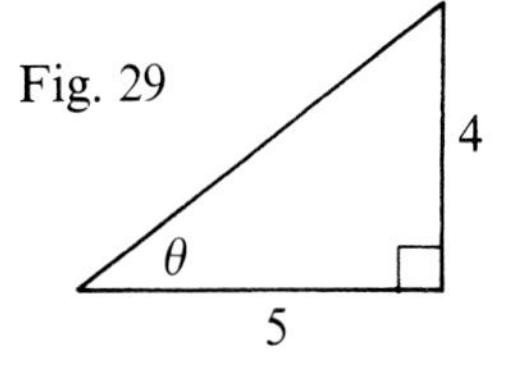

Fig. 29

3.

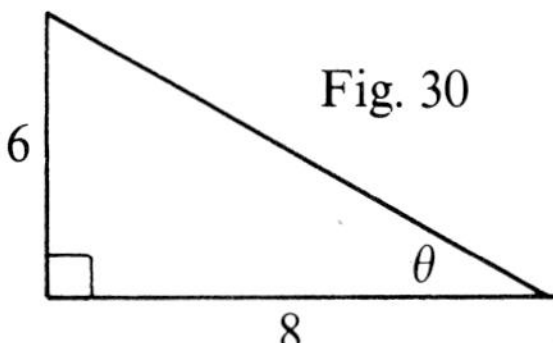

Fig. 30

4.

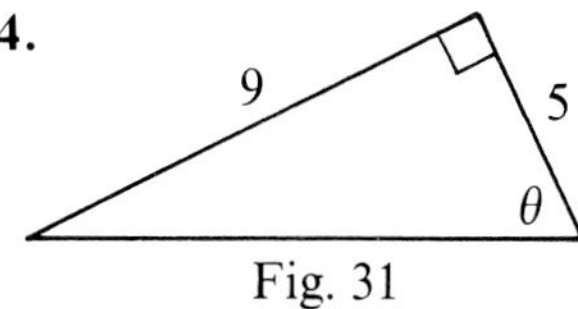

Fig. 31

5.

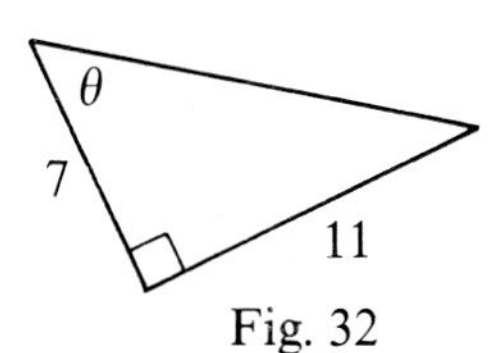

Fig. 32

6.

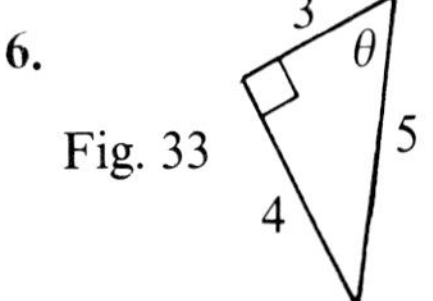

Fig. 33

7.

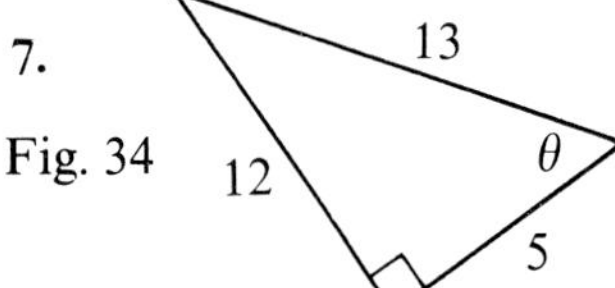

Fig. 34

8.

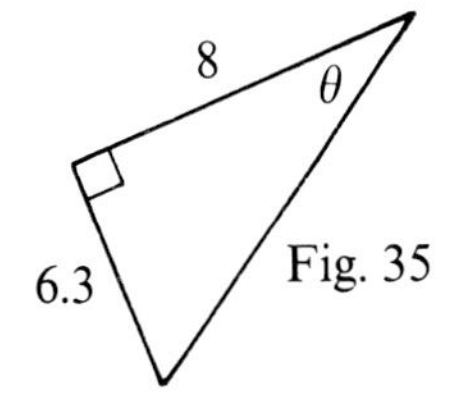

Fig. 35

9.

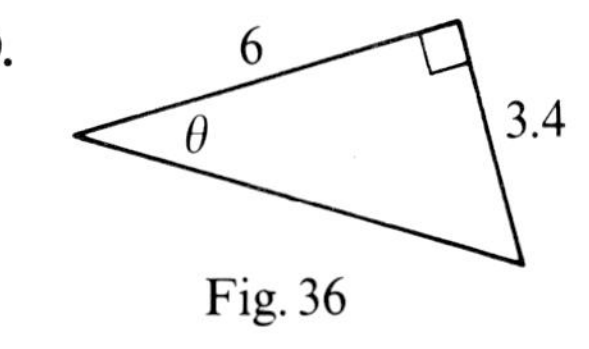

Fig. 36

Questions **10** to **19**: Give each answer to the nearest degree.

10. A rectangle has sides of 6 cm and 10 cm. Calculate the angle between
(i) a diagonal and a long side
(ii) the two diagonals.

11. The foot of a ladder is 2 m from a wall and it reaches a window ledge 3.6 m above the ground. Find the angle between the ladder and the ground.

12. On a flight of steps each step is 20 cm high and 30 cm deep. At what angle to the horizontal is the handrail?

13. The base of an isosceles triangle is 14 cm and the vertical height is 16 cm. Calculate a base angle.

14. What is the angle of elevation of the top of a cliff 82 m high at a point 60 m from it?

15.

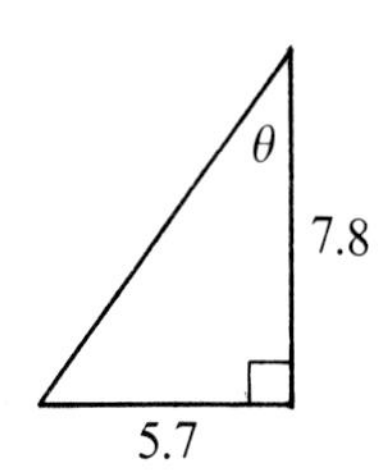

Calculate θ, using logarithms

Fig. 37

16.

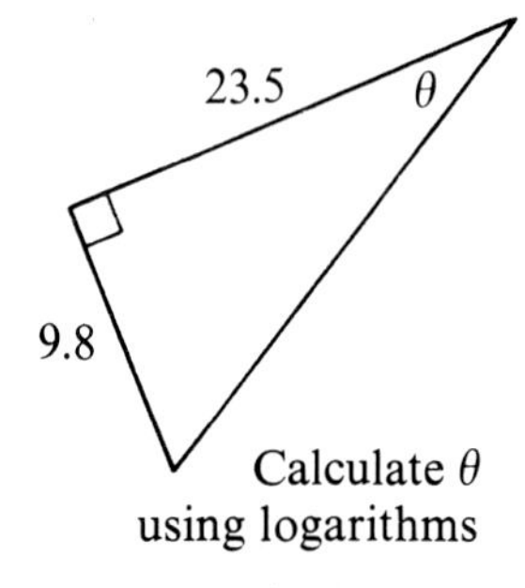

Calculate θ using logarithms

Fig. 38

17. The diagonals of a rhombus are 9.4 cm and 7.6 cm. Calculate the angles of the rhombus.

18. Calculate $\widehat{ABC}$.

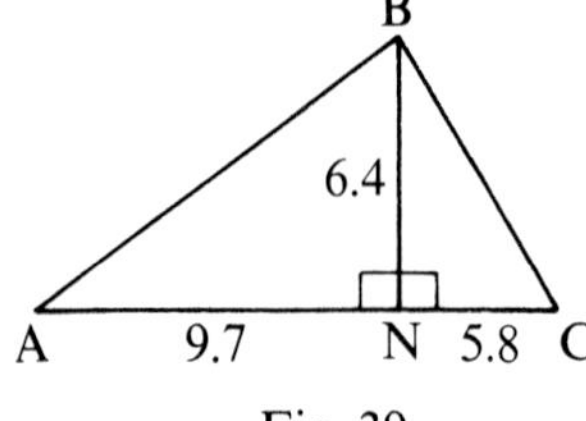

Fig. 39

19. On graph paper draw the graphs of $y = x$, $y = 2x$, $y = 3x$ and $y = 4x$. By constructing triangles, obtain the gradient of each line in the form n:1. (n is a whole number.) Using the tangent tables, find the acute angle each line makes with the x axis. Check by using a protractor.

12 · COMBINING TRANSFORMATIONS

In Mathematics we combine elements of the same sort to give another element of that sort. Here are some examples:

Numbers: $5 + 7 = 12$ and $3 \times 8 = 24$

Sets: $A \cup B = C$ and $A \cap B = D$

Matrices: $\begin{pmatrix} 3 & 5 \\ 2 & 1 \end{pmatrix} + \begin{pmatrix} 9 & 0 \\ 8 & 7 \end{pmatrix} = \begin{pmatrix} 12 & 5 \\ 10 & 8 \end{pmatrix}$

In this chapter we combine transformations.

Use a 45° set square or a triangle of the same shape made from card. Take the length of the equal sides as 1 unit.

EXAMPLE 1: *Let T_1 be a translation of 2 units and T_2 be a translation of 3 units in the same direction. Carry out T_1 on the triangle and then T_2.*

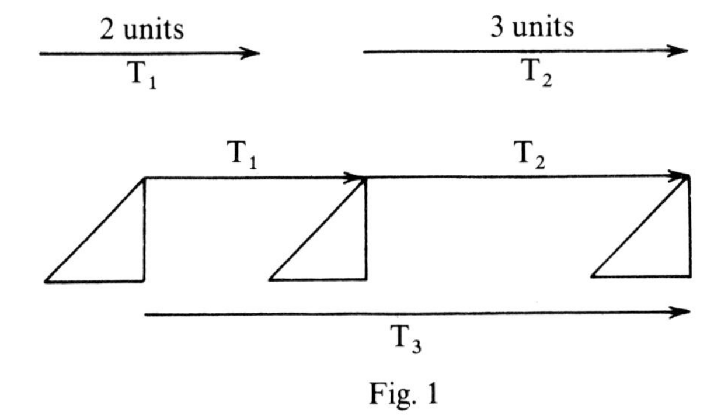

Fig. 1

Translations T_1 followed by T_2 have the same result as a single translation T_3 of 5 units in the same direction as T_1 and T_2. We write $T_3 = T_1 + T_2$.

EXAMPLE 2: *Carry out on the triangle the translations T_1 and T_2 shown in Fig. 2.*

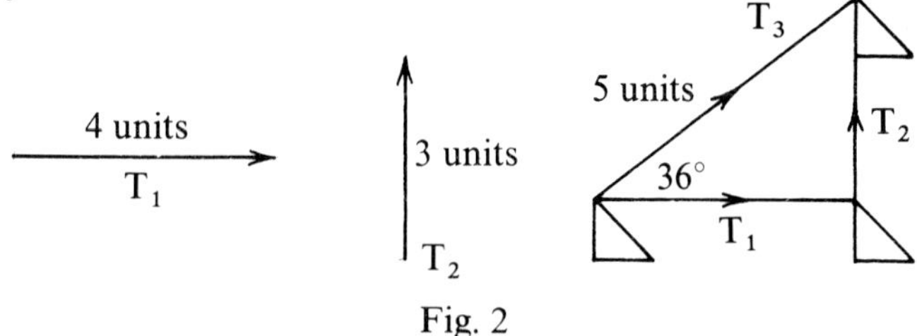

Fig. 2

Fig. 2 shows that the result is the same as for a single translation T_3 of 5 units in a direction of 36° to T_1. (Notice the 3, 4, 5 triangle.) Again we write $T_3 = T_1 + T_2$.

Exercise 49

1. Use a 45° set square, or a triangle of similar shape cut from card, to carry out the following pairs of translations. In each case show the single translation which has the same result.

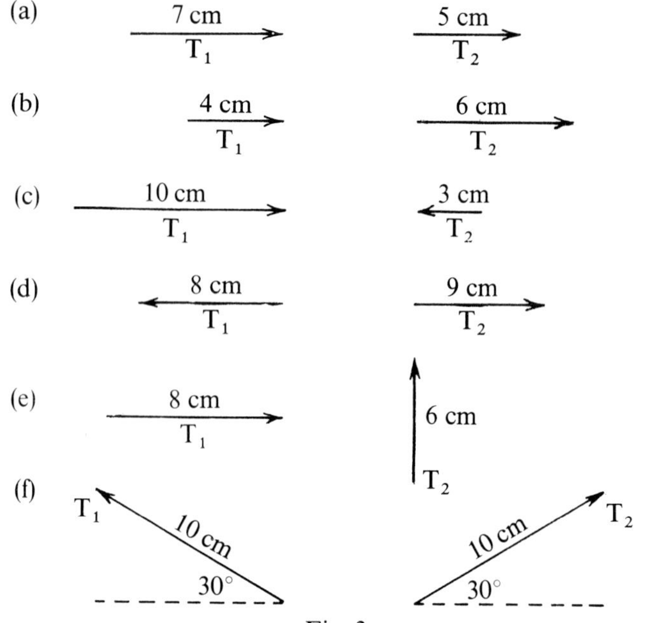

Fig. 3

2. In Question **1**, $T_1 + T_2$ was found by carrying out T_1 and then T_2. Now find $T_2 + T_1$ in each case by carrying out T_2 and then T_1. Do your results suggest that addition of translations is commutative?
That is, do they show that $T_1 + T_2 = T_2 + T_1$?

3. The T_1 and T_2 shown in Fig. 4 move an object the same distance but in opposite directions.

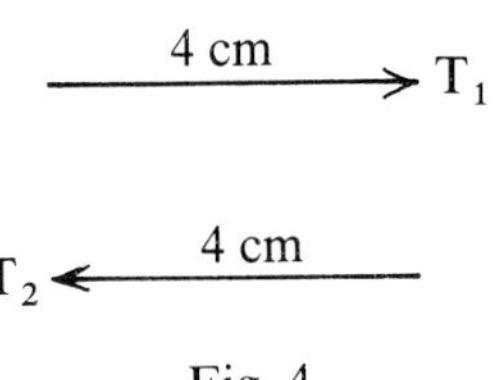

Fig. 4

We say that T_2 is the negative of T_1 and we write $T_2 = -T_1$. Draw the negatives of the translations shown in Fig. 5.

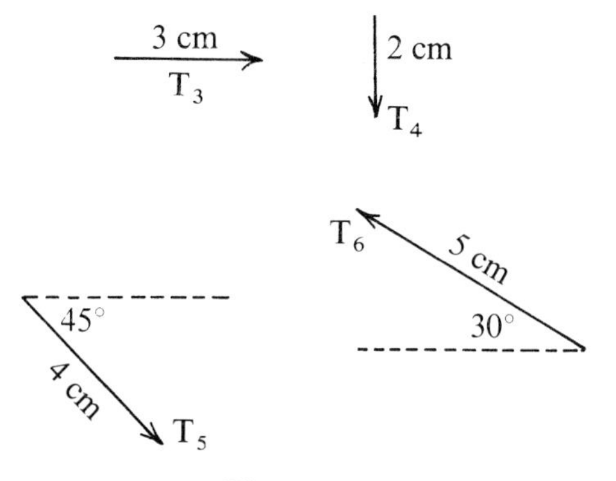

Fig. 5

4. T_1 followed by $(-T_4)$ can be written $T_1 + (-T_4)$ or more simply as $T_1 - T_4$. In Fig. 6, the single transformation which has the same result as T_1 followed by $(-T_4)$ is labelled $T_1 - T_4$. Draw diagrams to show:

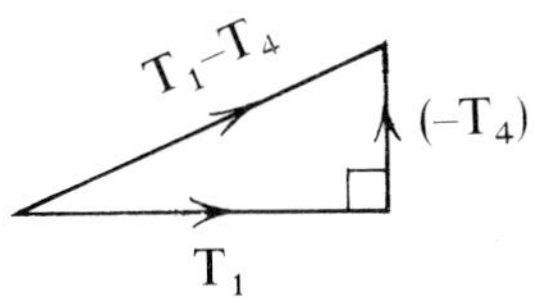

Fig. 6

(i) $T_2 - T_4$ (ii) $T_4 - T_3$ (iii) $T_5 - T_2$
(iv) $T_3 - T_6$ (v) $T_3 - T_5$ (vi) $T_6 - T_4$.

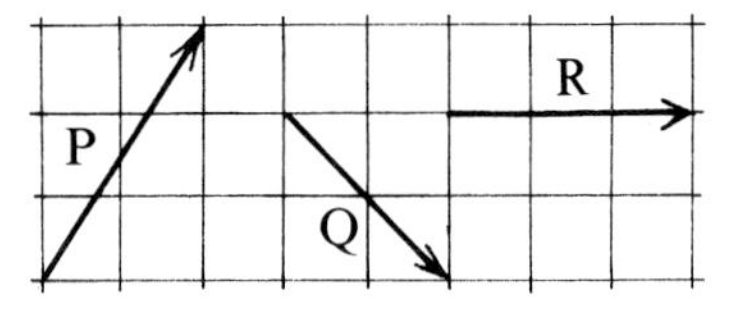

Fig. 7

5. Translations P, Q and R are shown in Fig. 7. On squared paper draw and label triangles, as in Fig. 6, for:

(i) P + Q (ii) Q − P (iii) R − Q
(iv) P + R (v) R − P (vi) Q − R.

COMBINING REFLECTIONS

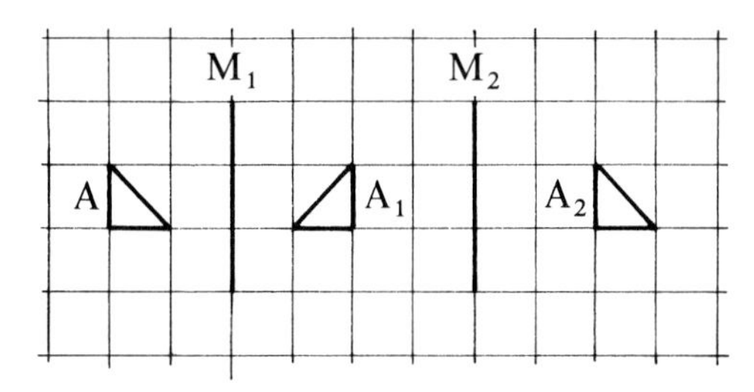

Fig. 8

In Fig. 8, M_1 and M_2 are parallel axes of reflection, distance 4 units apart. A is the object. A_1 is the image of A in M_1 and A_2 is the image of A_1 in M_2. The combined reflections move object A to position A_2. You can see that a single translation of 8 units to the right has the same result. Notice that the distance of 8 units is twice the distance between the axes of reflection, M_1 and M_2.

Exercise 50

1. On squared paper draw axes of reflection, M_1 and M_2, as in Fig. 8. Allow at least 11 squares to the left of M_1 and 7 to the right of M_2. Draw and label B_1, the image of A in M_2. Draw and label B_2, the image of B_1 in M_1. Check the following:
(i) B_2 is not in the same position as A_2.
(ii) A single translation of 8 units to the left has the same result.
(iii) This distance of translation is twice the distance between M_1 and M_2.

2. Copy Fig. 9 on squared paper, allowing at least 6 squares to the left of M_1 and 6 to the right of M_2. M_1 and M_2 are axes of reflection and A is the object.

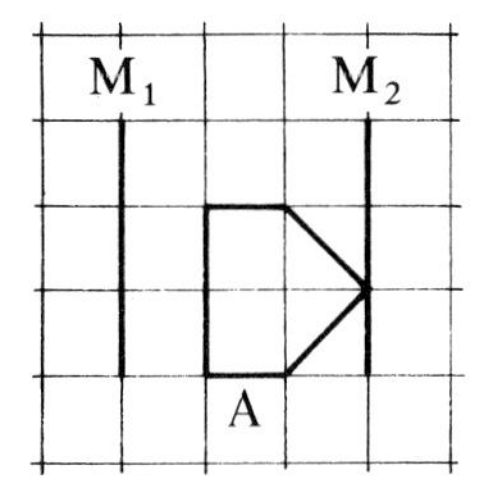

Fig. 9

Draw A_1, the image of A in M_1 and then draw A_2, the image of A_1 in M_2.

Using a different colour, draw B_1, the image of A in M_2 and then B_2, the image of B_1 in M_1.

Check the following statements:

(i) A can be moved to A_2 by a single translation.
(ii) A can be moved to B_2 by a single translation.
(iii) The distance of each translation is twice the distance between the mirrors.
(iv) A_2 and B_2 are not in the same place.

3. Copy Fig. 10 on squared paper, allowing at least 8 squares to the left of M_1 and 13 to the right of M_2. Repeat the work of Question **2** using this figure.

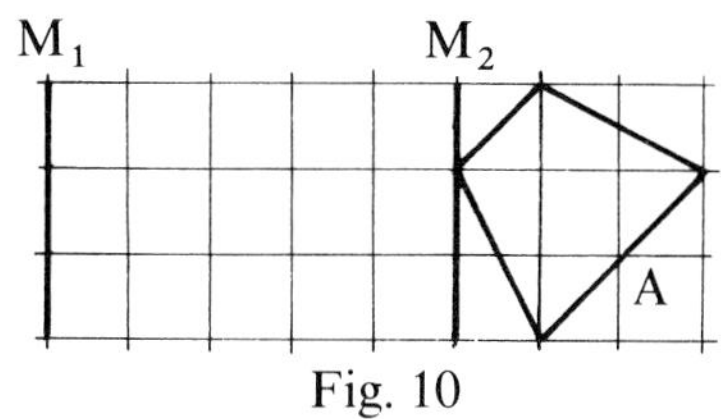

Fig. 10

Because the final images, A_2 and B_2 are not in the same place, the process of combining reflections is not commutative.

4. In Fig. 11, the axes of reflection, M_1 and M_2, are at 45° to each other. A_1 is the image of A in M_1 and A_2 is the image of A_1 in M_2. Copy the figure on squared paper and check that:

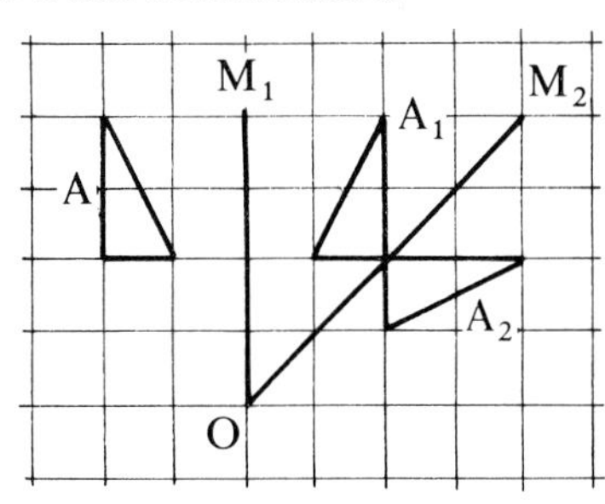

Fig. 11

(i) A can be moved to A_2 by a rotation about O,
(ii) the angle of rotation is twice the angle between the axes of reflection.

5. Make another copy of Fig. 11. Draw B_1, the image of A in M_2 and then B_2, the image of B_1 in M_1.
Check the following statements:
(i) The result is the same as for a rotation about O.
(ii) The angle of rotation is twice the angle between M_1 and M_2.
(iii) The position of B_2 is not the same as that of A_2 in Question **4.**

6. Copy Fig. 12 and draw:
(i) A_1, the image of A in M_1 and A_2, the image of A_1 in M_2.
(ii) B_1, the image of A in M_2 and B_2, the image of B_1 in M_1.

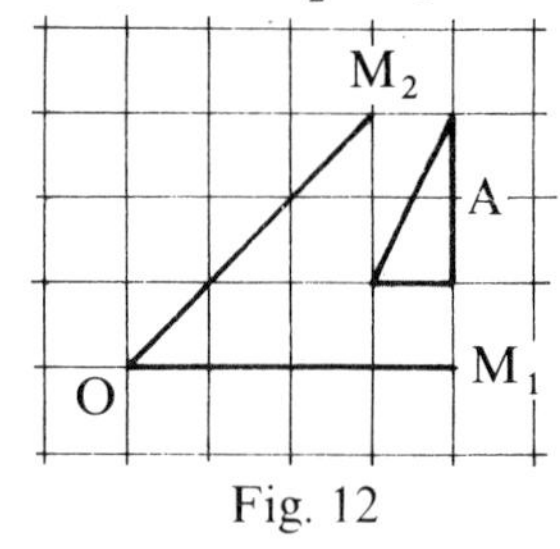

Fig. 12

Check that the statements of Question **5** apply here.

7. Copy Fig. 13 and draw A_1 and A_2 as for Question **6.** Check that A can be moved to A_2 by a rotation of 180° about O, the intersection of M_1 and M_2.

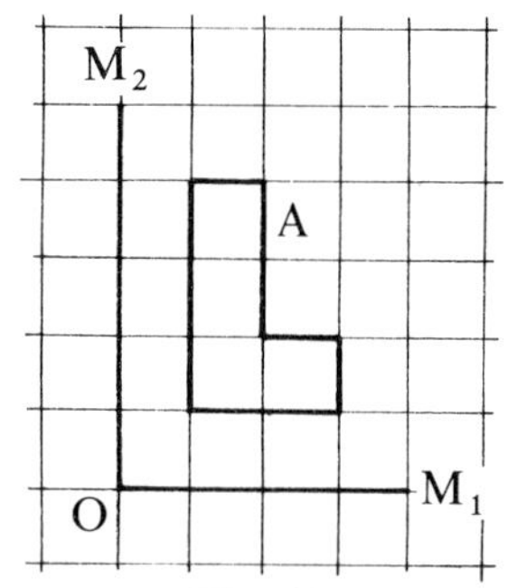

Fig. 13

8. Make another copy of Fig. 13 and draw images B_1 and B_2 as for Question **6.** You should find that A_2 and B_2 are in the same place. When the axes of reflection are at right angles to each other, the process of combining rotations is commutative.

9. Copy Fig. 14 and construct the images A_1, A_2, B_1 and B_2 as before. Check that:
(i) A_2 and B_2 occupy the same position, and
(ii) a rotation of 180° about O takes A to the same place.

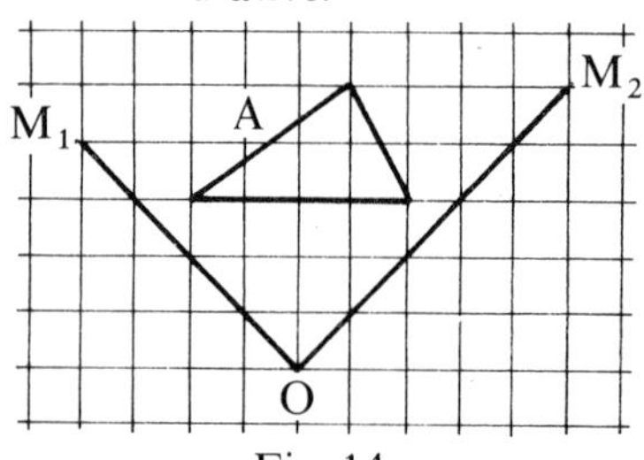

Fig. 14

SUMMARY

1. If one translation is followed by another, the result is the same as for a single translation. The process of combination is commutative, that is, $T_1 + T_2 = T_2 + T_1$.

2. If a reflection in one axis is followed by a reflection in a parallel axis, the result is the same as for a single translation. The distance of the translation is twice the distance between the axes.

3. If a reflection in one axis is followed by a reflection in an axis at $\theta°$ to the first, the result is the same as for a rotation through an angle of $2\theta°$ about the point of intersection of the axes.

Exercise 51

1. In Fig. 15, object A is translated to position A_1 and then image A_1 is rotated through 90° clockwise about P to position A_2. Copy the figure and use your compasses to check that A_2 can be obtained from A by a rotation through 90° about O.

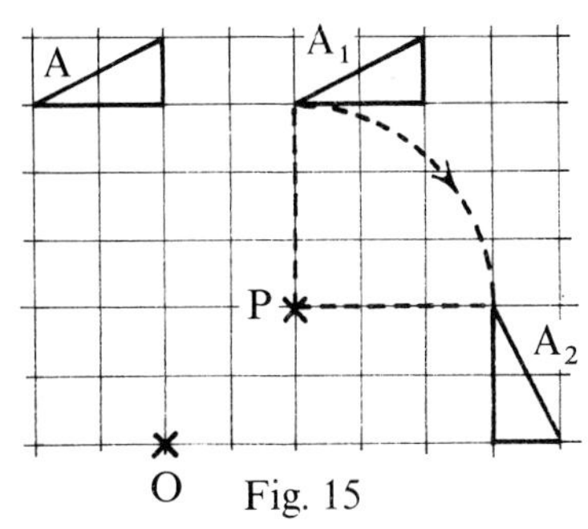

Fig. 15

Here a translation followed by a rotation has the same result as a rotation about a different point.

2. Copy Fig. 16. Draw A_1 as the image of A under a translation of 4 units to the right. Draw A_2 as the image of A_1 after a rotation through 90° clockwise about P. Use your compasses to check that A_2 can be obtained from A by a rotation through 90° about O.

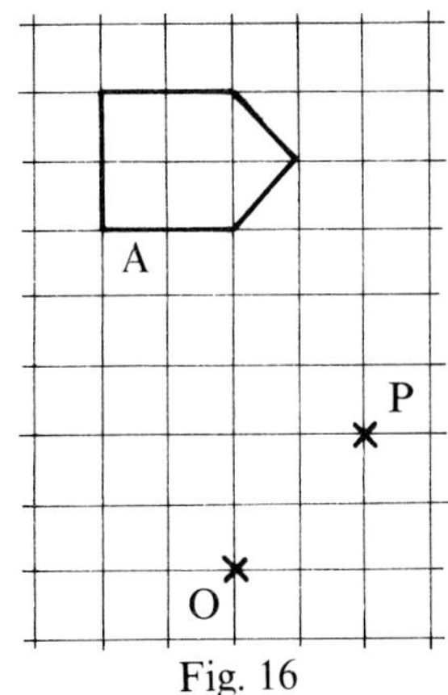

Fig. 16

3. In Fig. 17, object A is rotated through 90° clockwise about P to position A_1. Then A_1 is rotated through 90° anticlockwise about Q to position A_2. What single transformation would take A to A_2?

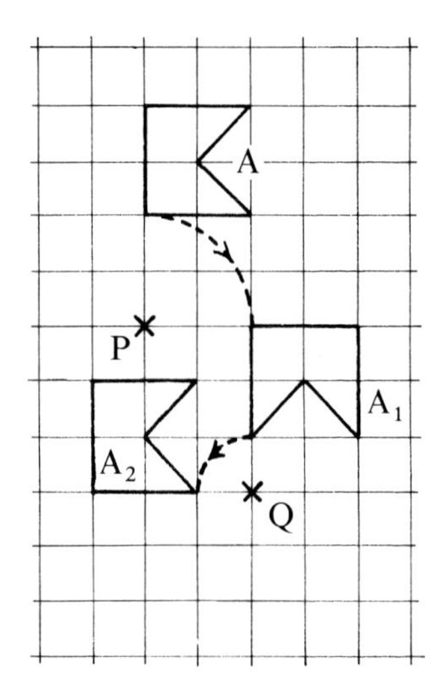

Fig. 17

4. Copy Fig. 18. A_1 is the image of A under an *anti-clockwise* rotation through 90° about P. Draw A_2, the image of A_1 under a *clockwise* rotation through 90° about Q. Check that A_2 is the image of A under a translation of 2 units to the ~~right~~. left.

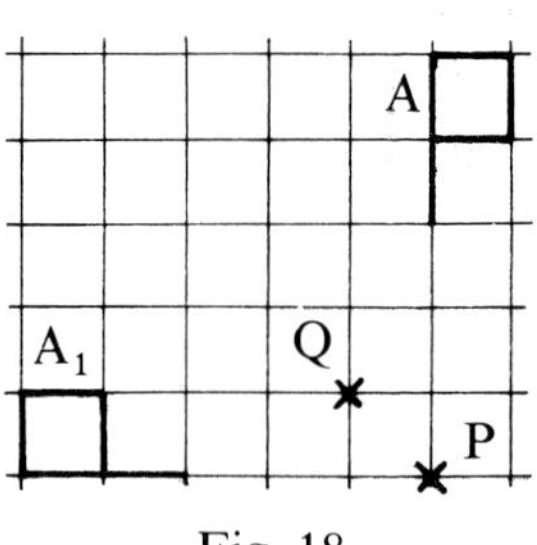

Fig. 18

5. Copy Fig. 19. Draw A_1 as the image of A under an *anticlockwise* rotation through 90° about P and then A_2 as the image of A_1 under a *clockwise* rotation through 90° about Q. What translation takes A to A_2?

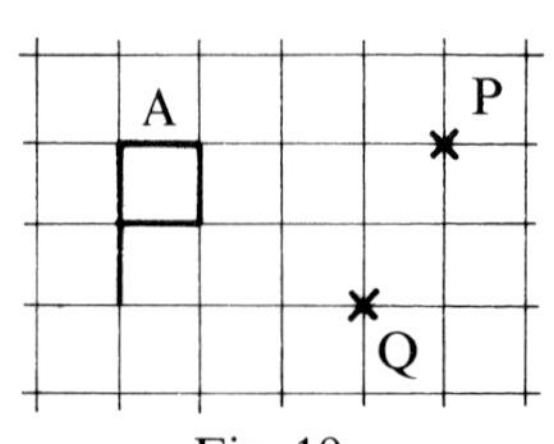

Fig. 19

6. In Fig. 20, O is the centre of enlargement. A_1B_1 is the image of AB under an enlargement with scale factor 3. A_2B_2 is the image of A_1B_1 with scale factor 2. Check that A_2B_2 is the image of AB with scale factor $2 \times 3 = 6$.

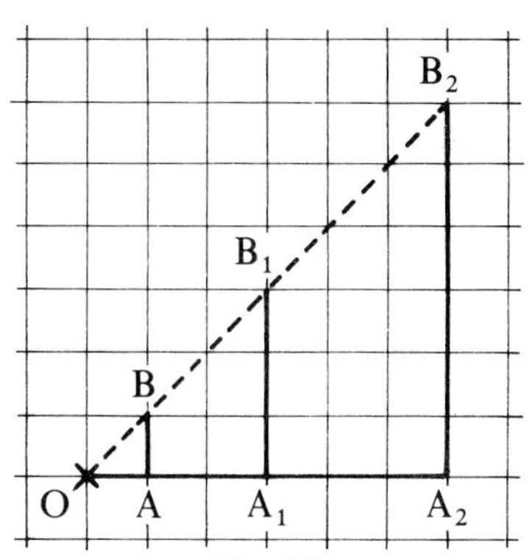

Fig. 20

(i) On squared paper draw object AB and centre of enlargement, O, as in Fig. 20. Draw A_1B_1 as the image of AB using scale factor 2 and then draw A_2B_2 as the image of A_1B_1 using scale factor 4. Check that A_2B_2 is the image of AB with scale factor 8.

(ii) Repeat (i) using a scale factor of 3 and then a scale factor of 3 again. What scale factor enlarges AB to A_2B_2?

7. Copy Fig. 21 on squared paper. $\triangle ABC$ is the object and O is the centre of enlargement. First draw $\triangle A_1B_1C_1$ as the image of $\triangle ABC$ with scale factor 4. Then draw $\triangle A_2B_2C_2$ as the image of $\triangle A_1B_1C_1$ with scale factor $\frac{1}{2}$.
What is the scale factor for enlarging $\triangle ABC$ to $\triangle A_2B_2C_2$?

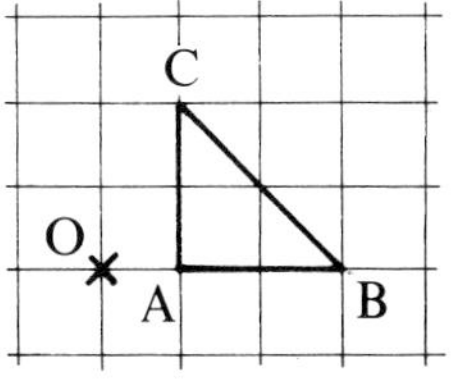

Fig. 21

8. Copy Fig. 22. With ABCD as object, O as centre of enlargement and a scale factor of $\frac{1}{3}$, draw image $A_1B_1C_1D_1$. With $A_1B_1C_1D_1$ as object, O as centre of enlargement and scale factor of 4, draw the image $A_2B_2C_2D_2$.

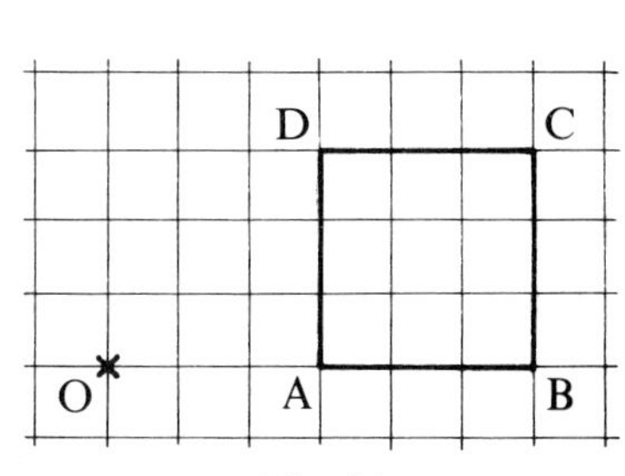

Fig. 22

What single enlargement changes ABCD to $A_2B_2C_2D_2$?

SUMMARY

1. A translation followed by a rotation is equivalent to a single rotation.
2. Two successive rotations through the same angle but in opposite directions are equivalent to a translation.
3. Two successive enlargements using centre O and scale factors p and q are equivalent to a single enlargement with centre O and scale factor pq.

13 · ALGEBRAIC PRODUCTS

By the distributive law

$$(a + b)p = ap + bp$$

Fig. 1 illustrates this statement.

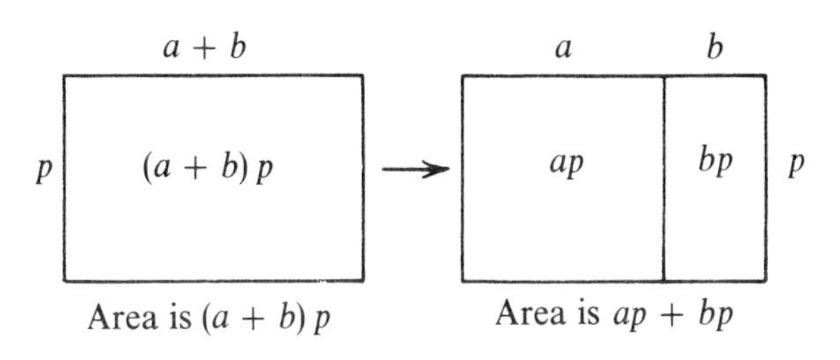

Fig. 1

If we now replace p with $(c + d)$, the statement becomes

$$\begin{aligned}(a + b)(c + d) &= a(c + d) + b(c + d)\\ &= ac + ad + bc + bd\end{aligned}$$

It is illustrated in Fig. 2.

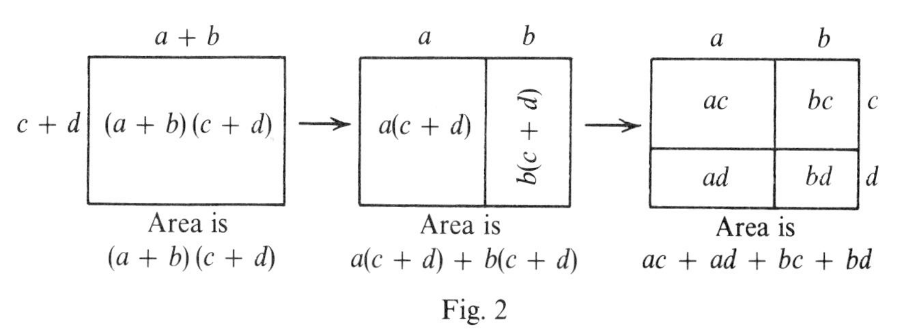

Fig. 2

Similarly, $(x + 5)q = xq + 5q$ and by replacing q with $(x + 3)$ we have

$$\begin{aligned}(x + 5)(x + 3) &= x(x + 3) + 5(x + 3)\\ &= x^2 + 3x + 5x + 15\\ &= x^2 + 8x + 15\end{aligned}$$

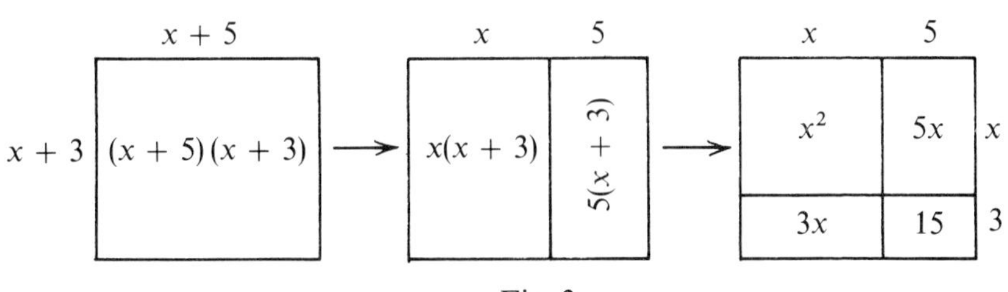

Fig. 3

We now have two ways of expanding a product such as $(3y + 4)(2y + 3)$.

METHOD 1: *Using a diagram.*

We draw a rectangle and divide it into four parts (Fig. 4). Then, as the area of the rectangle is equal to the sum of the four parts,

$$(3y + 4)(2y + 3) = 6y^2 + 17y + 12$$

	$3y$	4
$2y$	$6y^2$	$8y$
3	$9y$	12

Fig. 4

METHOD 2: *Using the distributive law.*

$$\begin{aligned}(3y + 4)(2y + 3) &= 3y(2y + 3) + 4(2y + 3)\\ &= 6y^2 + 9y + 8y + 12\\ &= 6y^2 + 17y + 12\end{aligned}$$

Exercise 52

1. Draw diagrams to illustrate the following statements:

(i) $(f + g)h = fh + gh$
(ii) $x(x + 7) = x^2 + 7x$

(iii) $(k + m)(g + h) = kg + kh + mg + mh$
(iv) $(x + 3)(y + 2) = xy + 2x + 3y + 6$
(v) $(p + 4)(p + 2) = p^2 + 6p + 8$.

2. (a) Use Fig. 5 to find the expansion of $(f + 6)(g + 4)$
(b) Write down a statement using Fig. 6.

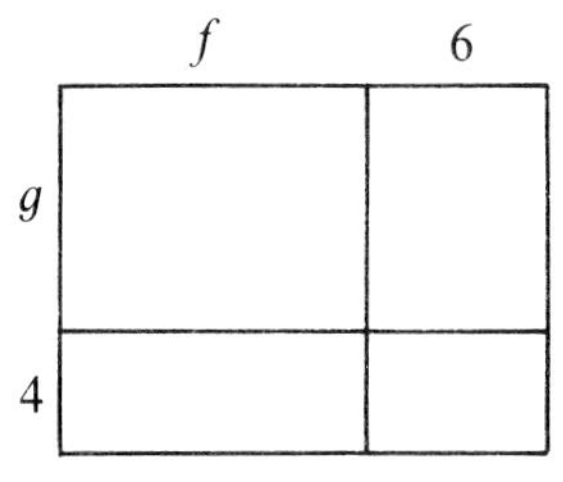

Fig. 5

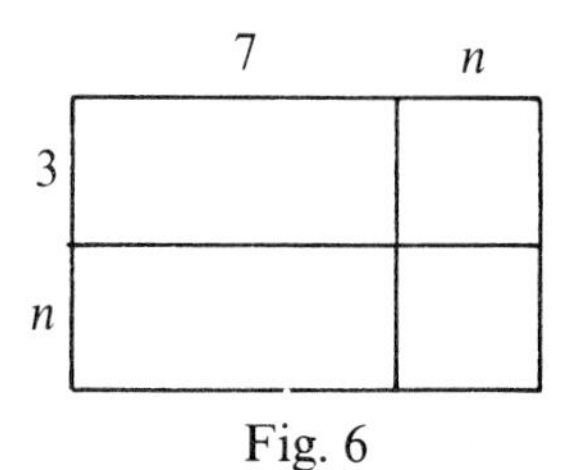

Fig. 6

3. Draw diagrams to obtain the expansions of:

(i) $(k + 3)(m + 2)$ (ii) $(n + 4)(p + 1)$
(iii) $(q + 5)(q + 8)$ (iv) $(u + 9)(u + 7)$
(v) $(4y + 3)(3y + 2)$.

4. *Without* drawing diagrams, obtain the expansions of:

(i) $3(a + 2)$ (ii) $b(c + 4)$
(iii) $(x + 1)(y + 3)$ (iv) $(m + 7)(n + 5)$
(v) $(p + q)(r + q)$ (vi) $(4 + k)(1 + n)$
(vii) $(h + 6)(h + 2)$ (viii) $(u + 3)(u + 1)$
(ix) $(5 + w)(2 + w)$.

5. Express without brackets:

(i) $8(2a + 3)$ (ii) $5b(3c + 7)$
(iii) $(3g + 2)(h + 6)$ (iv) $(m + 4)(5t + 3)$
(v) $(n + 5)(2n + 7)$ (vi) $(3u + 4)(2n + 1)$
(vii) $(4x + 1)(3x + 5)$ (viii) $(3 + 2y)(3 + 4y)$
(ix) $(1 + 5p)(7 + 2p)$.

6. (a) Use Fig. 7 to find the expansion of:
$(a + b + c)(x + y)$
(b) Draw diagrams to find the expansion of:
(i) $(f + g + 3)(f + 2)$
(ii) $(5 + n)(7 + k + n)$.

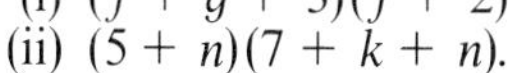

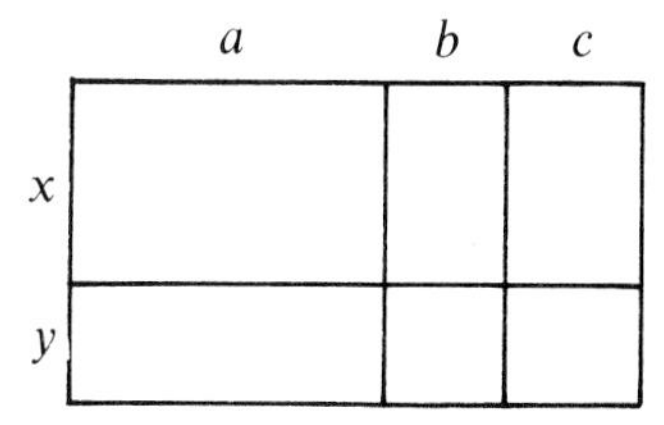

Fig. 7

BRACKETS WITH MINUS SIGNS

$$(a - b)\,p = ap - bp$$

Replacing p with $(c + d)$, this statement becomes

$$\begin{aligned}(a - b)(c + d) &= a(c + d) - b(c + d)\\ &= ac + ad - bc - bd\end{aligned}$$

Again, by replacing p with $(g - h)$ we have

$$\begin{aligned}(a - b)(g - h) &= a(g - h) - b(g - h)\\ &= ag - ah - bg + bh\end{aligned}$$

(Remember that just as $(-5) \times (-2) = +10$, so $(-b) \times (-h) = +bh$.)

Here is another example:

$$(2 - 3x)\,q = 2q - 3xq$$

Hence
$$\begin{aligned}(2 - 3x)(5 - 2x) &= 2(5 - 2x) - 3x(5 - 2x)\\ &= 10 - 4x - 15x + 6x^2\\ &= 10 - 19x + 6x^2\end{aligned}$$

Exercise 53

1. (i) Expand $(g + 5)p$ and hence $(g + 5)(k - 2)$
(ii) Expand $(h - 3)p$ and hence $(h - 3)(n + 4)$
(iii) Expand $(7 - w)q$ and hence $(7 - w)(2 - w)$.

2. Expand:

(i) $(x - 2)(y + 1)$ (ii) $(x + 2)(y - 1)$
(iii) $(x - 2)(y - 1)$ (iv) $(u - 7)(w - 3)$
(v) $(h - 5)(h - 2)$ (vi) $(4 - k)(3 - k)$
(vii) $(6 + m)(3 - m)$ (viii) $(1 - d)(f + 4)$
(ix) $(c - 2)(5 - c)$.

3. Expand:

(i) $(2x - 1)(x + 4)$ (ii) $(y - 3)(4y + 3)$
(iii) $(3p + 1)(2u - 1)$ (iv) $(5a - 1)(2b - 3)$
(v) $(1 - 3c)(2 - 5c)$ (vi) $(4 + 3d)(1 - d)$
(vii) $(2f - 5)(1 + f)$ (viii) $(3g + 1)(2g - 3)$
(ix) $(3 - n)(2n + 3)$.

4. $q(c + d) = qc + qd$. Replacing q with $(a + b)$ we get
$(a + b)(c + d) = (a + b)c + (a + b)d = ac + bc + ad + bd$.
Is this the same result as was obtained on page 123?
Expand the following *both* ways:
(i) $(a + 3)(b + 4)$ (ii) $(c - 5)(c - 7)$
(iii) $(6 - d)(2 + f)$ (iv) $(k + 1)(m - 3)$.

5. (i) Expand $(x + 2)(x + 5)$ and hence solve the equation
$(x + 2)(x + 5) = x^2 + 9x$.
(ii) Solve $(y - 3)(y + 4) + 2y = y^2$.
(iii) Solve $(5 - u)(u - 3) = 9 - u^2$.
(iv) Solve $(6 + w)(3 - w) + w(w - 2) = 0$.

6. Solve:
(i) $(x + 3)(x - 2) = (x + 1)(x - 1)$
(ii) $(y + 2)(y - 1) = (y + 5)(y - 2)$
(iii) $(4p + 1)(p + 2) = (4p - 5)(p + 10)$
(iv) $(5n + 1)(n + 2) = (5n - 7)(n + 7)$.

7. (i) Expand $(x + 3)(x + 3)$. Notice that it is $(x + 3)^2$. Illustrate it with a diagram like Fig. 4.
(ii) Expand $(a + b)^2$ and draw a diagram to illustrate it.
(iii) Expand $(y + 5)^2$ and $(p + 7)^2$
(iv) Expand $(8 + t)^2$ and $(1 + 6w)^2$.

8. (i) Expand $(x - 5)(x - 5)$
(ii) Expand $(y - 4)^2$ and $(w - 7)^2$
(iii) Expand $(3 - p)^2$ and $(a - b)^2$.

9. Expand:
(i) $(a + 5)(a - 5)$ (ii) $(b - 3)(b + 3)$
(iii) $(2 + c)(2 - c)$ (iv) $(7 - d)(7 + d)$
(v) $(x - y)(x + y)$ (vi) $(3f + 1)(3f - 1)$
Comment on your answers and on the given expressions.
Write down, without any working, the expansion of:
$(g + 8)(g - 8)$ and of $(h - 9)(h + 9)$.

10. Using the discovery made in Question **9**, copy and complete:
(i) $n^2 - 25 = (n + 5)(\ldots\ \ldots)$
(ii) $p^2 - 64 = (\ldots\ \ldots)(\ldots\ \ldots)$
(iii) $t^2 - 16 = (\ldots\ \ldots)(\ldots\ \ldots)$
(iv) $36 - y^2 = (\ldots\ \ldots)(\ldots\ \ldots)$

11. (i) Copy and complete:
$(a + 1)(b + 1) = a(b + 1) + 1(\ldots\ \ldots) = ab + \ldots + \ldots + 1$

and so

$$31 \times 51 = (30 + 1)(50 + 1) = 30(50 + 1) + 1(\ldots\ \ldots)$$
$$= 1500 + \ldots + \ldots + 1 = \ldots.$$

(ii) Copy and complete:

$$(k - 2)(n - 1) = k(n - 1) - 2(\ldots\ \ldots) = kn - \ldots - \ldots + 2$$

and so

$$98 \times 29 = (100 - 2)(30 - 1) = 100(30 - 1) - 2(\ldots\ \ldots)$$
$$= 3000 - 100 - \ldots + 2 = \ldots.$$

Use similar methods for:

(iii) 63×22 (iv) 29×19 (v) 81×59
(vi) 0.9×1.2 (vii) 98×28 (viii) 3.2×7.3.

12. $65^2 = (70 - 5)(60 + 5)$. By expanding the brackets show that $65^2 = 70 \times 60 + 25 = 4225$.

Use this idea to write down the squares of 35, 75, 85, 4.5 and 2.5.

IMPORTANT IDENTITIES

Here are two statements which have the same appearance:

$$\text{A:}\ 7(x + 3) - 4(x + 5) = 2x + 4$$
$$\text{B:}\ 5(x + 2) - 3(x + 1) = 2x + 7$$

They are really quite different. Removing the brackets on the left-hand side of A, we get,

$$7x + 21 - 4x - 20 = 3x + 1$$

so that for A to be true,

$$3x + 1 = 2x + 4$$

and

$$x = 3$$

Removing the brackets on the left-hand side of B, we get,

$$5x + 10 - 3x - 3 = 2x + 7.$$

This is eactly the same as the right-hand side and hence B is true for every value of x.

A is an *equation*: B is an *identity*.

An equation is true for just certain values of the letter or letters in it: an identity is true for all values of the letter or letters.

We now obtain three important identities which should be memorised.

$$\begin{aligned}(a+b)(a+b) &= a(a+b)+b(a+b)\\ &= a^2+ab+ba+b^2\\ &= a^2+ab+ab+b^2\\ &= a^2+2ab+b^2\end{aligned}$$

Writing $(a+b)(a+b)$ as $(a+b)^2$ we have the identity

$$(a+b)^2 = a^2+2ab+b^2 \qquad \textbf{(1)}$$

If $a=5$ and $b=3$, this becomes

$$\begin{aligned}(5+3)^2 &= 5^2+2\times 5\times 3+3^2\\ 64 &= 25+30+9\end{aligned}$$

Similarly we can obtain

$$(a-b)^2 = a^2-2ab+b^2 \qquad \textbf{(2)}$$

and

$$(a+b)(a-b) = a^2-b^2 \qquad \textbf{(3)}$$

EXAMPLE 1: *Expand:*

(i) $(x+7)^2$ (ii) $(3y-5)^2$ (iii) $(5-8\text{p})(5+8\text{p})$.

(i) Replacing a by x and b by 7 in (**1**),

$$\begin{aligned}(x+7)^2 &= x^2+2\times x\times 7+7^2\\ &= x^2+14x+49\end{aligned}$$

(ii) Replacing a by $3y$ and b by 5 in (**2**),

$$\begin{aligned}(3y-5)^2 &= (3y)^2-2\times(3y)\times 5+5^2\\ &= 9y^2-30y+25\end{aligned}$$

(iii) Replacing a by 5 and b by $8p$ in (**3**),

$$\begin{aligned}(5-8p)(5+8p) &= 5^2-(8p)^2\\ &= 25-64p^2\end{aligned}$$

EXAMPLE 2: *Find the value of*: (i) 73^2 (ii) 9.9^2 using (**1**) and (**2**).

(i) Putting $a=70$ and $b=3$ in (**1**),

$$\begin{aligned}73^2 = (70+3)^2 &= 70^2+2\times 70\times 3+3^2\\ &= 4900+420+9\\ &= 5329\end{aligned}$$

(ii) Putting $a = 10$ and $b = 0.1$ in (**2**),

$$9.9^2 = (10 - 0.1)^2 = 10^2 - 2 \times 10 \times 0.1 + 0.1^2$$
$$= 100 - 2 + 0.01$$
$$= 98.01$$

Exercise 54

1. Expand:

(i) $(a + 3)^2$ (ii) $(b + 5)^2$ (iii) $(c + 8)^2$
(iv) $(d - 4)^2$ (v) $(e - 7)^2$ (vi) $(f - 10)^2$
(vii) $(4 - d)^2$ (viii) $(5 + b)^2$ (ix) $(10 - f)^2$.

2. Expand:

(i) $(g + 5)(g - 5)$ (ii) $(h + 3)(h - 3)$ (iii) $(k - 7)(k + 7)$
(iv) $(4 - m)(4 + m)$ (v) $(6 - n)(6 + n)$ (vi) $(1 - p)(1 + p)$
(vii) $(3r + 1)(3r - 1)$ (viii) $(5t - 1)(5t + 1)$ (ix) $(1 - 4x)(1 + 4x)$

3. Expand:

(i) $(3x + y)^2$ (ii) $(c + 7d)^2$ (iii) $(f + 2g)^2$
(iv) $(h - 4k)^2$ (v) $(5m - n)^2$ (vi) $(3p - q)^2$
(vii) $(r + 8t)(r - 8t)$ (viii) $(v - 5w)(v + 5w)$ (ix) $(2x - y)(2x + y)$

4. Expand:

(i) $(cd - 3)(cd + 3)$ (ii) $(5 - fg)^2$ (iii) $(2h + 3m)^2$
(iv) $(3k - 2n)^2$ (v) $(5p + 2r)^2$ (vi) $(t^3 + 1)^2$
(vii) $(x^3 + 1)(x^3 - 1)$ (viii) $(3v + 4w)(3v - 4w)$ (ix) $(2 - y^2)(2 + y^2)$.

5. Use the identities (**1**) and (**2**) from page 129 to calculate

(i) 21^2 (ii) 39^2 (iii) 52^2 (iv) 48^2
(v) 201^2 (vi) 199^2 (vii) 9.8^2 (viii) 19.7^2.

6. If $a = 60$ and $b = 7$, the identity (**3**) becomes $(60 + 7)(60 - 7) = 60^2 - 7^2$ from which:

$$67 \times 53 = 3600 - 49 = 3551$$

(i) By putting $a = 90$ and $b = 3$, find the value of 93×87.
(ii) By putting $a = 4$ and $b = 0.2$, find the value of 4.2×3.8.

Use the method to calculate:

(iii) 58×42 (iv) 8.4×7.6 (v) 205×195 (vi) 0.32×0.28.

7. Use the method of Example 2 before this exercise to calculate, correct to 2 decimal places:

(i) 5.03^2 (ii) 1.05^2 (iii) 2.04^2 (iv) 4.01^2.

8. Use identity (**2**) to calculate, correct to 2 decimal places:
(i) 4.97^2 (ii) 2.95^2 (iii) 3.98^2 (iv) 6.99^2.

9. Copy and complete:
(i) $a^2 + 10a + 25 = (a + \ldots)^2$
(ii) $b^2 - 14b + 49 = (b - \ldots)^2$
(iii) $c^2 - 18c + 81 = (\ldots \quad \ldots)^2$
(iv) $d^2 - 20d + 100 = (\ldots \quad \ldots)^2$
(v) $f^2 - 25 = (\ldots \quad \ldots)(\ldots \quad \ldots)$
(vi) $g^2 - 81 = (\ldots \quad \ldots)(\ldots \quad \ldots)$.

10. The following statements are wrong. Think what changes should be made to the right-hand side of each and write down your corrected statement.
(i) $(p + 6)^2 = p^2 + 6p + 36$
(ii) $(v - 11)^2 = v^2 + 22v - 121$
(iii) $(x + 12)^2 = x^2 + 12x + 24$
(iv) $(4 - 3y)^2 = 16 - 12y + 9y^2$
(v) $(w + 5)^2 = w^2 + 25$.

11. Copy and complete:
(i) $9 + 6h + h^2 = (\ldots \quad \ldots)^2$
(ii) $1 - 12k + 36k^2 = (\ldots \quad \ldots)^2$
(iii) $m^6 - 10m^3 + 25 = (\ldots \quad \ldots)^2$
(iv) $1 - n^2 = (\ldots \quad \ldots)(\ldots \quad \ldots)$
(v) $9p^2 - 1 = (\ldots \quad \ldots)(\ldots \quad \ldots)$
(vi) $q^2 - 16 = (\ldots \quad \ldots)(\ldots \quad \ldots)$
(vii) $4t^2 - 12t + 9 = (\ldots \quad \ldots)^2$
(viii) $9x^2 - 30xy + 25y^2 = (\ldots \quad \ldots)^2$.

12. Two of the following are identities and two are equations. State which are identities and solve the equations.
A: $3(x + 1) + 2(x + 4) = 5(x + 2) + 1$
B: $4(x + 1) + 3(x + 2) = 5(x + 4) - 2$
C: $(x + 1)(x + 6) = (x + 3)^2$
D: $(x + 5)^2 = (x + 2)^2 + 3(2x + 7)$.

13. Here is an identity:
$(x + 1)^3 = x^3 + 3x^2 + 3x + 1$
(i) Test it by putting $x = 3$ and $x = 5$.
(ii) Use it to solve the equation
$x(x^2 + 3x + 5) = (x + 1)^3$.

14. Show that the following are identities:

(i) $a(a - 1) = a^2 - a$
(ii) $(b + 3)^2 - 6b = b^2 + 9$

(iii) $(x + y)^2 - 4xy = (x - y)^2$
(iv) $(n + 1)^2 = (n + 5)(n - 3) + 16$
(v) $(h - 4)^2 = (4 - h)^2$
(vi) $(p^2 + 1)^2 = (p^2 - 1)^2 + (2p)^2$

The identity of part (vi) was used in Exercise **24**, Question **9**, to obtain some Pythagorean triads.

14 · ALGEBRAIC FACTORS

TYPE $ab + ac = a(b + c)$

Some questions of this type were worked in Book 2.
Here are some further examples.

$$5h + 20 = 5(h + 4)$$
$$p^2 - pt = p(p - t)$$
$$8bc - 12fc = 4c(2b - 3f)$$
$$6x + 3y - 15z = 3(2x + y - 5z)$$

Exercise 55

Factorise:

1. $3a + 3b$ **2.** $cd - ce$ **3.** $2f + 10$
4. $6 - 3g$ **5.** $hm + 2m$ **6.** $n^2 + 5n$
7. $4p - p^2$ **8.** $r^2 + r$ **9.** $t - t^2$
10. $5ab + 5ac$ **11.** $2cd - 6ce$ **12.** $10f^2 - 5f$
13. $3g + 6g^2$ **14.** $10hk - 15hm$ **15.** $p^6 + p^4$
16. $t^5 - 2t^2$ **17.** $ab + ac + ad$ **18.** $4f - 6g + 10h$.

TYPE $a^2 - b^2 = (a + b)(a - b)$

This identity was established on page 129.

Putting $a = x$ and $b = 3$, $x^2 - 9 = (x + 3)(x - 3)$
Putting $a = 5y$ and $b = 7$, $25y^2 - 49 = (5y + 7)(5y - 7)$
Similarly, $81 - 4p^2 = 9^2 - (2p)^2 = (9 + 2p)(9 - 2p)$

$$q^6 - 100 = (q^3)^2 - 10^2 = (q^3 + 10)(q^3 - 10)$$
$$\tfrac{4}{9} - r^2 = (\tfrac{2}{3})^2 - r^2 = (\tfrac{2}{3} + r)(\tfrac{2}{3} - r)$$

Exercise 56

Factorise:

1. $x^2 - 25$ **2.** $y^2 - 36$ **3.** $c^2 - 81$
4. $64 - d^2$ **5.** $9 - f^2$ **6.** $1 - g^2$

7. $9h^2 - 4$ **8.** $49 - 9k^2$ **9.** $36m^2 - 1$
10. $1 - 4n^2$ **11.** $p^2 - \frac{1}{9}$ **12.** $q^2 - \frac{4}{25}$
13. $r^2 - 2\frac{1}{4}$ **14.** $1 - t^6$ **15.** $w^{16} - 25$.
16. Copy and complete: $2x^2 - 50 = 2(x^2 - \ldots) = 2(x + \ldots)(x - \ldots)$.

Factorise:

17. $3y^2 - 12$ **18.** $5z^2 - 45$ **19.** $ab^2 - 9a$
20. $c^3 - cd^2$ **21.** $49f - 25f^3$.

22. $86^2 - 14^2 = (86 + 14)(86 - 14) = 100 \times 72 = 7200$. Use this method to calculate:

(i) $73^2 - 27^2$ (ii) $65^2 - 55^2$ (iii) $103^2 - 3^2$ (iv) $133^2 - 67^2$
(v) $0.7^2 - 0.3^2$ (vi) $3.7^2 - 1.7^2$ (vii) $12.7^2 - 7.3^2$ (viii) $0.9^2 - 0.8^2$.

23. Use the identity to calculate:

(i) $\sqrt{(17^2 - 15^2)}$ (ii) $\sqrt{(41^2 - 40^2)}$
(iii) $\sqrt{(5.2^2 - 4.8^2)}$ (iv) $\sqrt{(2.9^2 - 2.1^2)}$.

MORE PRODUCTS

$$\begin{aligned}(x + 5)(x + 2) &= x(x + 2) + 5(x + 2)\\ &= x^2 + 2x + 5x + 10\\ &= x^2 + 7x + 10\end{aligned}$$

Notice how the terms x^2, $7x$ and 10 arise;

x^2: $(x + 5)(x + 2)$ $2x$, $5x$: $(x + 5)(x + 2)$ 10: $(x + 5)(x + 2)$

The product of the FIRST in each bracket gives x^2;

the product of the OUTSIDES added to the product of the INSIDES gives $7x$;

the product of the LAST in each bracket gives 10.

If we take the first letter of each of the words FIRST, OUTSIDES, INSIDES, LAST we have FOIL. This may help you to remember the order.

Using this knowledge we can shorten the working out of products.

$9x$, $6x$:

$$(x + 6)(x + 9) = x^2 + 9x + 6x + 54 = x^2 + 15x + 54$$

$$(y - 5)(y + 3) = y^2 + 3y - 5y - 15 = y^2 - 2y - 15$$

(with arrows indicating $3y$ and $-5y$)

Exercise 57

Expand:

1. (i) $(x + 3)(x + 4)$ (ii) $(y + 1)(y + 6)$ (iii) $(p + 10)(p + 5)$
(iv) $(q + 7)(q + 4)$ (v) $(k + 6)(k + 11)$ (vi) $(n + 4)(n + 9)$

2. (i) $(a - 3)(a - 5)$ (ii) $(b - 2)(b - 1)$ (iii) $(c - 4)(c - 3)$
(iv) $(d - 6)(d - 2)$ (v) $(e - 1)(e - 9)$ (vi) $(f - 5)(f - 8)$.

3. (i) $(g + 4)(g - 2)$ (ii) $(h + 7)(h - 4)$ (iii) $(k - 7)(k + 4)$
(iv) $(m - 9)(m + 3)$ (v) $(n - 5)(n + 8)$ (vi) $(p + 5)(p - 8)$
(vii) $(q + 4)(q - 4)$ (viii) $(t - 1)(t + 1)$ (ix) $(u - 6)(u + 3)$.

4. (a) (i) If there are + signs in both brackets, what signs occur in the product?
(ii) If there are − signs in both brackets, what signs occur in the product?

(b) Copy and complete:

(i) $x^2 + 6x + 5 = (x + \ldots)(x + \ldots)$
(ii) $y^2 - 4y + 3 = (y - \ldots)(y - \ldots)$
(iii) $n^2 + 8n + 7 = (\ldots \quad \ldots)(\ldots \quad \ldots)$
(iv) $p^2 - 5p + 6 = (\ldots \quad \ldots)(\ldots \quad \ldots)$.

5. (a) Look back at Question **3.** If one bracket has a + sign and the other has a − sign, what can you say about the sign of the last term in the product?

(b) Copy and complete:

(i) $f^2 + 2f - 3 = (f + \ldots)(f - \ldots)$
(ii) $g^2 - 4g - 5 = (g + \ldots)(g - \ldots)$
(iii) $h^2 + 6h - 7 = (\ldots \quad \ldots)(\ldots \quad \ldots)$
(iv) $m^2 - m - 6 = (\ldots \quad \ldots)(\ldots \quad \ldots)$.

Suppose that we wish to factorise $x^2 + 7x + 6$. As there are two + signs we know that the factors have the form $(x + \ldots)(x + \ldots)$. We need two numbers having a product of 6. We could use 3 and 2, or 6 and 1.

$$(x + 3)(x + 2) = x^2 + 2x + 3x + 6 = x^2 + 5x + 6$$

and $(x + 6)(x + 1) = x^2 + x + 6x + 6 = x^2 + 7x + 6$

Thus the required factorisation is $(x + 6)(x + 1)$.

Now consider $x^2 - 12x + 20$. As the middle term has a $-$ sign and the last term has a $+$ sign, the factors are of the form $(x - \ldots)(x - \ldots)$. They might be $(x - 1)(x - 20)$, $(x - 2)(x - 10)$ or $(x - 4)(x - 5)$. Which is correct?

Exercise 58

1. Factorise:

(i) $x^2 + 4x + 3$ (ii) $y^2 + 3y + 2$ (iii) $p^2 + 12p + 11$
(iv) $q^2 - 3q + 2$ (v) $r^2 - 6r + 5$ (vi) $t^2 - 14t + 13$.

2. Two of the following cannot be factorised. State which and factorise the other two.

(i) $x^2 + 8x + 7$ (ii) $x^2 + 10x + 7$
(iii) $x^2 - 6x + 7$ (iv) $x^2 - 8x + 7$.

3. (a) Expand $(x + 2)(x + 5)$ and $(x + 1)(x + 10)$. Which is the factorisation of $x^2 + 7x + 10$?

(b) Expand $(y + 1)(y + 12)$, $(y + 2)(y + 6)$ and $(y + 3)(y + 4)$. Which is the factorisation of $y^2 + 8y + 12$?

(c) Expand $(u + 1)(u + 24)$, $(u + 2)(u + 12)$, $(u + 3)(u + 8)$ and $(u + 4)(u + 6)$. Which is the factorisation of $u^2 + 10u + 24$?

4. (a) If $18 = a \times b$, state the possible pairs of values for a and b. (Positive integers only.) Hence factorise $x^2 + 11x + 18$ and $x^2 + 9x + 18$.

(b) State the pairs of positive integers which satisfy $30 = c \times d$. Hence factorise $p^2 - 11p + 30$ and $q^2 - 13q + 30$.

(c) Do the same for $16 = e \times f$, $n^2 + 8n + 16$, $k^2 + 10k + 16$.

(d) Do the same for $36 = g \times h$, $m^2 - 13m + 36$, $t^2 - 37t + 36$.

5. Factorise where possible:

(i) $a^2 + 7a + 12$ (ii) $b^2 + 8b + 12$ (iii) $c^2 + 9c + 12$
(iv) $d^2 - 6d + 8$ (v) $e^2 - 7e + 8$ (vi) $f^2 - 9f + 8$
(vii) $g^2 + 5g + 6$ (viii) $h^2 + 5h + 15$ (ix) $k^2 + 8k + 15$.

6. (a) Copy and complete $y^2 + \ldots y + 11$ so that it *cannot* be factorised.

(b) Copy and complete $p^2 + \ldots p + 2$ so that it *can* be factorised, and state its factors.

(c) Copy and complete $k^2 - \ldots k + 15$ in two different ways so that it can be factorised. State the factors in each case.

EXAMPLE: *Factorise* $x^2 - x - 6$

Because the last term has a $-$ sign, the brackets have the form $(x + \ldots)(x - \ldots)$.

The possibilities are:

$$(x+6)(x-1),\quad (x+3)(x-2),\quad (x+2)(x-3) \quad \text{and} \quad (x+1)(x-6).$$

Expand the brackets to find which is correct.

Exercise 59

1. Factorise:

(i) $a^2 - 2a - 3$ (ii) $b^2 + 2b - 3$ (iii) $c^2 + 6c - 7$
(iv) $d^2 - 6d - 7$ (v) $e^2 - 10e - 11$ (vi) $f^2 - 4f - 5$
(vii) $g^2 - g - 2$ (viii) $h^2 - 3h - 4$ (ix) $k^2 + 5k - 6$.

2. If $x^2 \ldots x - 10 = (\ldots \;\; \ldots)(\ldots \;\; \ldots)$, state the possible pairs of brackets and expand each pair. Hence factorise:

(i) $x^2 - 3x - 10$ (ii) $y^2 + 3y - 10$
(iii) $u^2 + 9u - 10$ (iv) $w^2 - 9w - 10$.

3. Factorise:

(i) $k^2 - k - 12$ (ii) $m^2 + m - 12$ (iii) $n^2 - 4n - 12$
(iv) $p^2 + 4p - 12$.

What other two quadratic expressions beginning with x^2 and ending with -12 can be factorised?

4. Factorise, where possible:

(i) $a^2 - 2a - 8$ (ii) $b^2 - 7b - 8$ (iii) $c^2 - 9c - 8$
(iv) $d^2 + 2d - 8$.

5. Factorise, where possible:

(i) $f^2 - 3f - 18$ (ii) $g^2 - 4g - 18$ (iii) $h^2 - 7h - 18$
(iv) $k^2 - 17k - 18$ (v) $m^2 - 12m - 18$ (vi) $n^2 + 3n - 18$.

6. Factorise, where possible:

(i) $p^2 - 8p - 9$ (ii) $r^2 - 9$ (iii) $u^2 - 6u + 9$
(iv) $w^2 - 10w + 9$ (v) $x^2 + 9$ (vi) $y^2 + 6y + 9$.

$$(5 - n)(3 + n) = 15 + 5n - 3n - n^2 = 15 + 2n - n^2$$

$$(x + 4y)(x - 7y) = x^2 - 7xy + 4xy - 28y^2 = x^2 - 3xy - 28y^2$$

$$(8 - p^3)(2 + p^3) = 16 + 8p^3 - 2p^3 - p^6 = 16 + 6p^3 - p^6.$$

Exercise 60

1. Expand:

(i) $(2 + a)(6 - a)$ (ii) $(3 - b)(7 + b)$ (iii) $(5 - c)(10 + c)$
(iv) $(6 + d)(1 - d)$ (v) $(12 + e)(3 - e)$ (vi) $(9 - f)(2 + f)$.

2. Copy and complete:

(i) $14 - 5g - g^2 = (\ldots + g)(\ldots - g)$
(ii) $10 - 9h - h^2 = (\ldots + h)(\ldots - h)$
(iii) $15 + 2k - k^2 = (\ldots \quad \ldots)(\ldots \quad \ldots)$
(iv) $8 - 7m - m^2 = (\ldots \quad \ldots)(\ldots \quad \ldots)$.

3. Expand:

(i) $(a - 4b)(a + 6b)$ (ii) $(c - 3d)(c - 7d)$ (iii) $(e^3 + 1)(e^3 - 5)$
(iv) $(2 + f^2)(7 + f^2)$ (v) $(3g - 1)(4g + 1)$ (vi) $(7h - n)(3h - n)$.

4. Copy and complete:

(i) $p^2 + 4pq - 5q^2 = (p + \ldots)(p - \ldots)$
(ii) $10u^2 - 3uv - v^2 = (\ldots + v)(\ldots - v)$
(iii) $x^2 - 6xy + 5y^2 = (x - \ldots)(x - \ldots)$
(iv) $10g^2 + 7gh + h^2 = (\ldots + h)(\ldots + h)$.

5. Factorise:

(i) $6 + 5x + x^2$ (ii) $10 - 7y + y^2$ (iii) $8 - 2u - u^2$
(iv) $12 + w - w^2$ (v) $8 - 9p + p^2$ (vi) $12 + 4r - r^2$.

6. Factorise:

(i) $1 - 5a + 4a^2$ (ii) $1 + 2b - 3b^2$ (iii) $c^2 - 3cd - 10d^2$
(iv) $3f^2 - 4fg + g^2$ (v) $h^4 + 6h^2 + 5$ (vi) $1 - k^3 - 6k^6$.

7. Factorise:

(i) $n^2 - 5n$ (ii) $k^2 + 8k$ (iii) $6p^2 - p$
(iv) $10q + q^2$ (v) $x^3 + 4x^2$ (vi) $3y^2 - 6y$.

8. Factorise, where possible:

(i) $16h^2 - 17h + 1$ (ii) $16k^2 + 1$ (iii) $16m^2 - 1$
(iv) $16p^2 + 8p + 1$ (v) $16x^2 - 8x - 1$ (vi) $16y^2 - 6y - 1$.

9. (a) Copy and complete:

(i) $a^3 + 3a^2 - 10a = a(a^2 \ldots\ldots) = a(a \quad \ldots)(a \quad \ldots)$
(ii) $b^4 - 81 = (b^2 + \ldots)(b^2 - \ldots) = (b^2 + \ldots)(b + \ldots)(b - \ldots)$.

(b) Factorise:

(i) $c^3 + 5c^2 + 6c$ (ii) $d^4 - 6d^3 - 7d^2$ (iii) $f^4 - 16$
(iv) $5g^2 - 10g - 15$ (v) $3h - 15h^2 + 12h^3$ (vi) $1 - n^4$.

15 · TOPOLOGY

Geology means the study of the earth: biology means the study of living things: topology means the study of place or position. *Topos* is a Greek word meaning 'place'. Sometimes topology is called rubber sheet geometry.

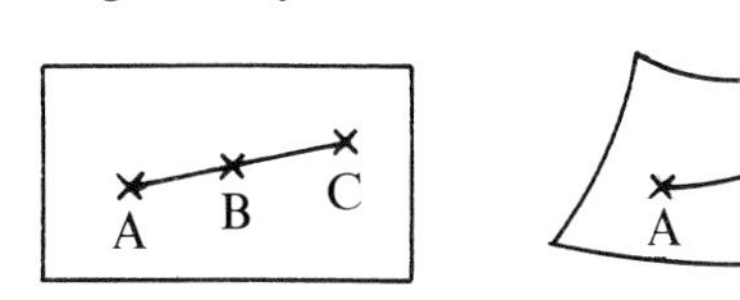

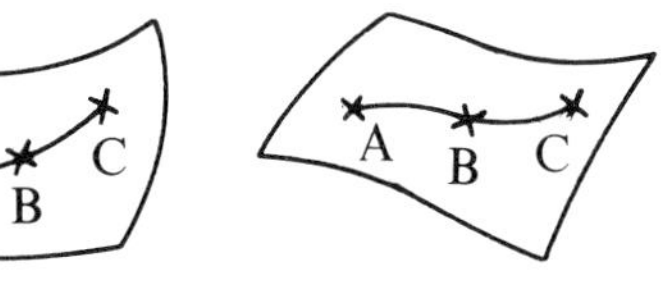

Fig. 1

Imagine a rectangular rubber sheet stretched in different directions (without folding or cutting or tearing). The straight line ABC becomes curved. Its length and direction change, but the order of the points A, B, C is still the same and the curve remains unbroken. We say that ABC is a topological line.

Exercise 61

1. Which of these are topological lines?

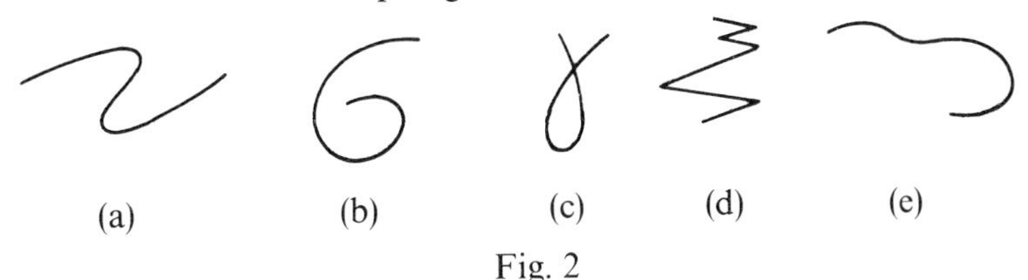

Fig. 2

2. Imagine the circle in Fig. 3 drawn on a rubber sheet. Draw two more diagrams to show the possible distortion of its shape when the rubber is stretched in different ways.

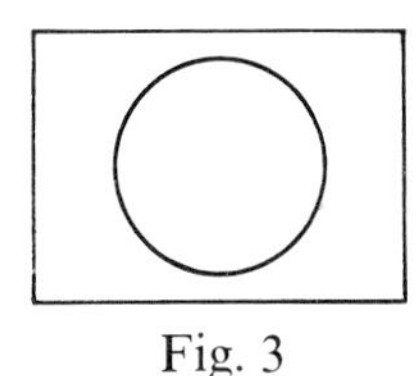

Fig. 3

3. In topology a circle is a *simple closed curve.* It has an inside region, an outside region and an unbroken boundary. Which of these figures are simple curves?

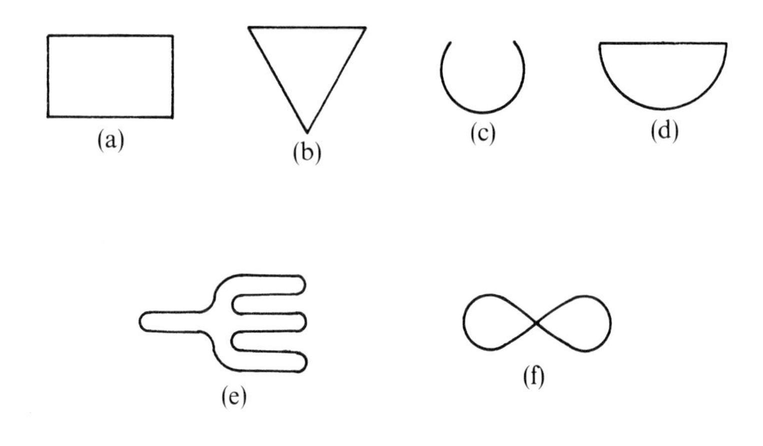

Fig. 4

4. In Fig. 5, P is a point inside a circle and Q is a point outside the circle. Two paths from P to Q are shown. One path cuts the boundary once and the other cuts it three times.

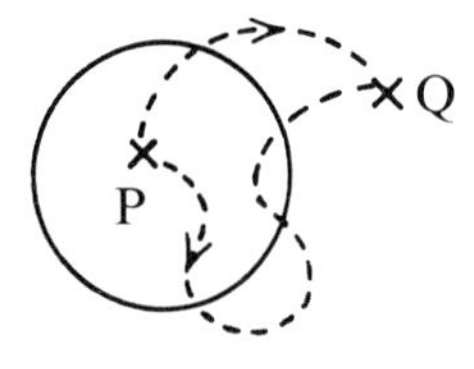

Fig. 5

Draw a circle and mark a point P inside it and a point Q outside it. Draw paths from P to Q cutting the circle:

(i) 3 times (ii) 5 times (iii) 7 times.

Is it possible to draw paths which cut the circle an even number of times such as 2 or 4?

5. Fig. 6 shows a simple closed curve. Notice that R is inside the curve and W is outside. In Question **4** we found that a path from a point inside to a point outside cuts the curve an odd number of times.

How many times does the path from R to W cut the curve? How many time does a path from S to V cut the curve? Is S inside or outside the curve? Use this method to find whether T is inside or outside the curve.

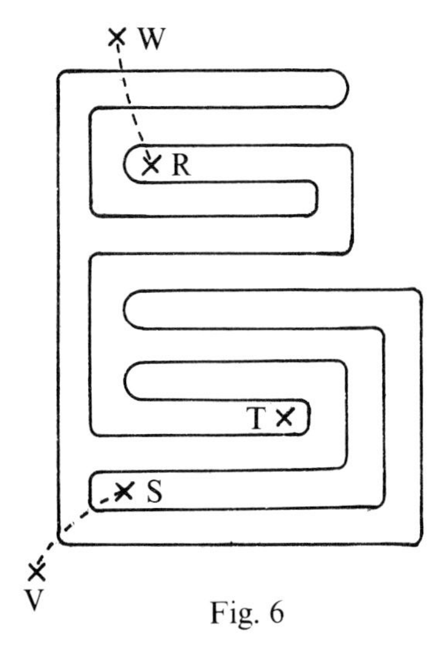

Fig. 6

6. The lines drawn in Fig. 7 divide the plane into 5 regions. State the number of regions into which the drawings in Fig. 8 divide the plane. Always count the outside as a region.

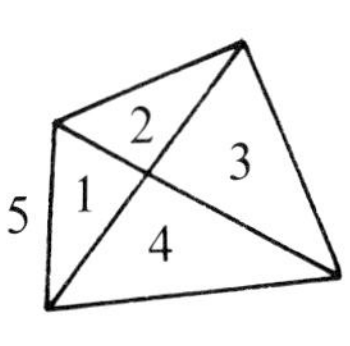

Fig. 7

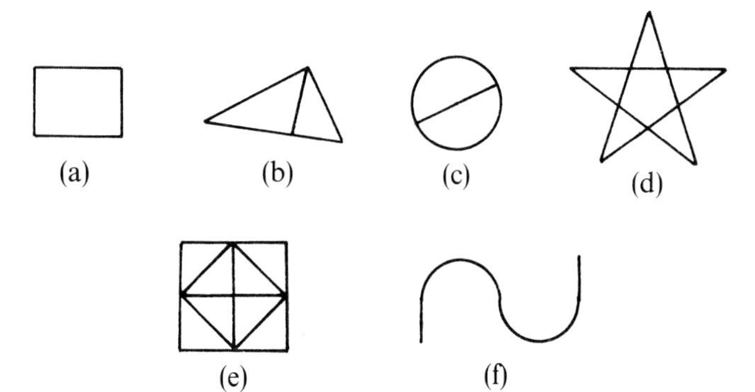

Fig. 8

7. Draw separate figures which divide the plane into:
(i) 1 (ii) 2 (iii) 3 (iv) 6 regions.

8. Which of the shapes in Fig. 9 are topologically the same. That is, which can be distorted into each other without tearing or folding?

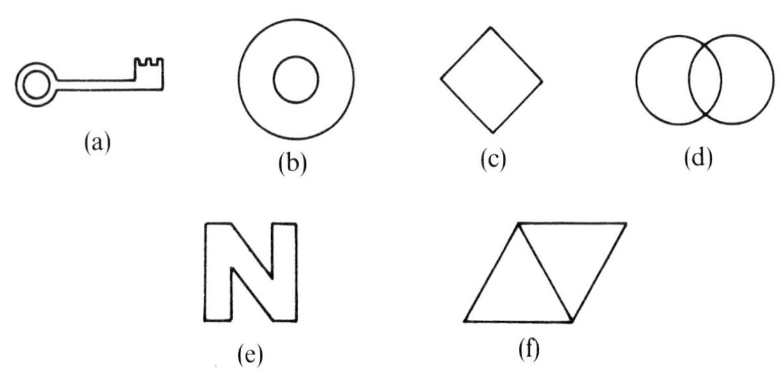

Fig. 9

9. The four parts of Fig. 10 are maps of countries. Copy each map and colour it so that no two countries with a common boundary have the same colour. What is the least number of colours you need?

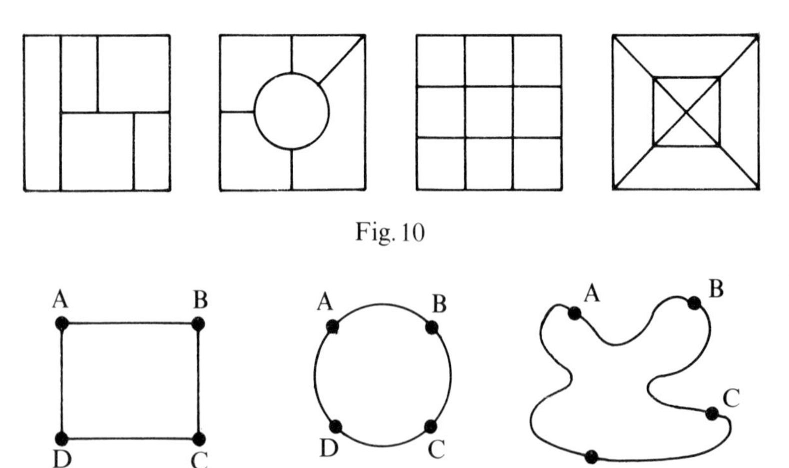

Fig. 10

Fig. 11

The three parts of Fig. 11 are topologically the same. At each of the points A, B, C and D two paths or *arcs* meet. The points A, B, C, and D are called *nodes*.

A figure consisting of nodes, arcs and regions is called a *network*.

The *order* of a node is the number of arcs meeting at that node. Here are some examples:

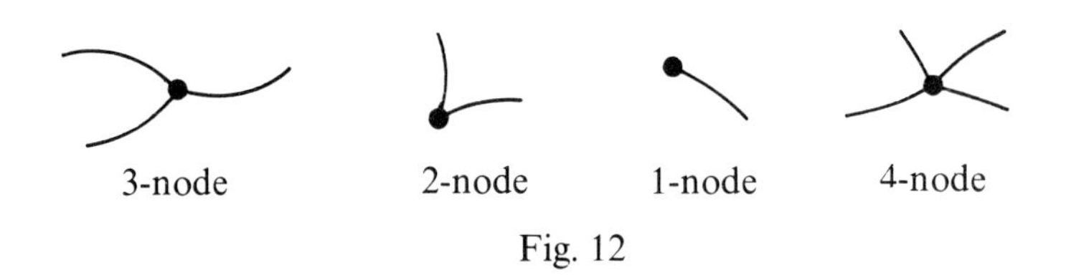

Fig. 12

Nodes where an even number (2, 4, 6, ...) of arcs meet are called *even nodes*. *Odd nodes* are junctions where an odd number of arcs meet.

Exercise 62

1. State the number of odd and even nodes in each of the networks in Fig. 13.

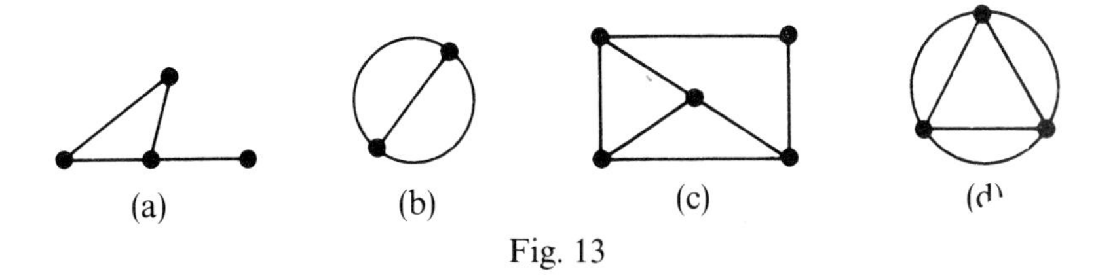

(a) (b) (c) (d)

Fig. 13

2. For each network in Fig. 13 state:
   (i) the number of arcs (ii) the number of regions.

3. Draw a network with:
   (i) three nodes each of order 2.
   (ii) two nodes each of order 4.
   (iii) three nodes each of order 1 and one node of order 3.

4. For each of the networks in Fig. 14 find the value of $N + R - A$ where N is the number of nodes, R is the number of regions and A is the number of arcs. What do you find?

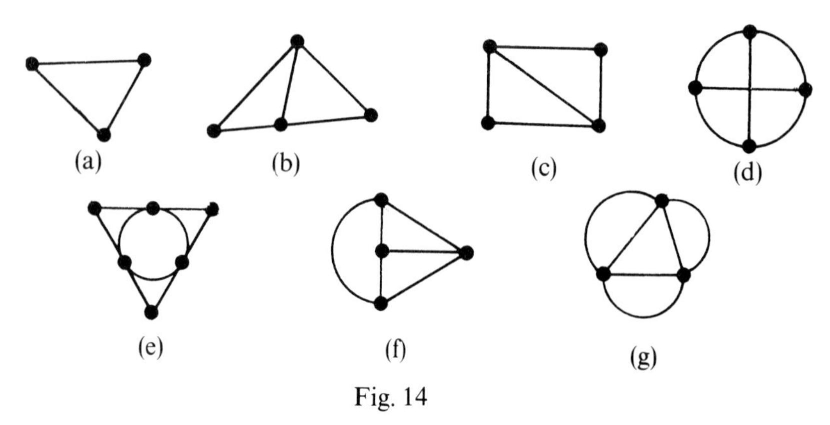

Fig. 14

5. Invent two networks of your own and check that $N + R - A = 2$.

TRAVERSABLE NETWORKS

Copy the network in Fig. 15. Starting at A it is possible to go over each arc once without taking your pencil off the paper. Do this with a coloured pencil by going from A to B and then to C, D, A and C as shown by the arrows in Fig. 16.

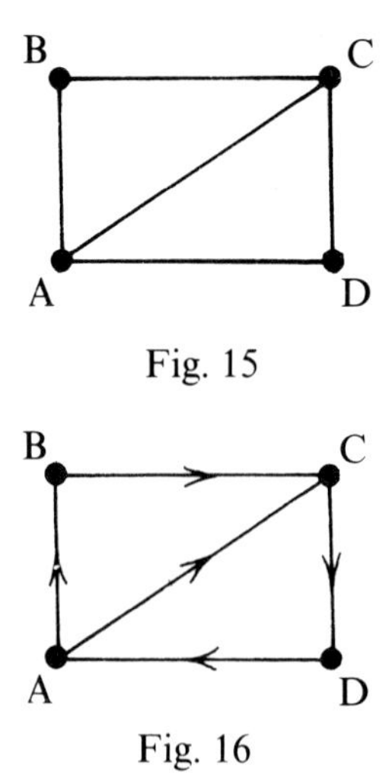

Fig. 15

Fig. 16

Make another copy of the network and go over it with a coloured pencil using the route A–C–B–A–D–C. On another copy try a different route from A. Notice that you are allowed to pass through a node more than once but you must not go over any arc more than once. Copy the network again and try to go over it by starting at B. You will find that it is not possible.

If it is possible to go over every arc of a network once without taking the pencil off the paper, as for the network of Fig. 15, then the network is said to be *traversable*.

Exercise 63

1. Copy each of the following networks. Try to trace them as described above. Which are traversable? In each case say whether or not you end at the point where you started.

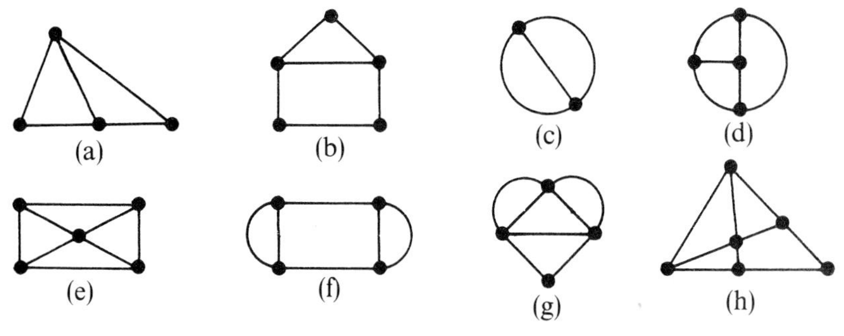

Fig. 17

2. Invent a network which is traversable and also one which is not traversable.

3. A table has been started below. Copy and complete it for the networks of Questions **1** and **2**

Network	*Number of even nodes*	*Number of odd nodes*	*Traversable*
1(a)	2	2	yes
1(b)	3		

Check that a network is traversable if it has:
(a) two odd nodes or (b) all even nodes.

16 · SINES AND COSINES

In Chapter 11, we used the tangent ratio to solve problems on right-angled triangles. Each problem involved the two sides next to the right angle, but did not involve the hypotenuse, the side opposite the right angle. In this section we shall use the hypotenuse.

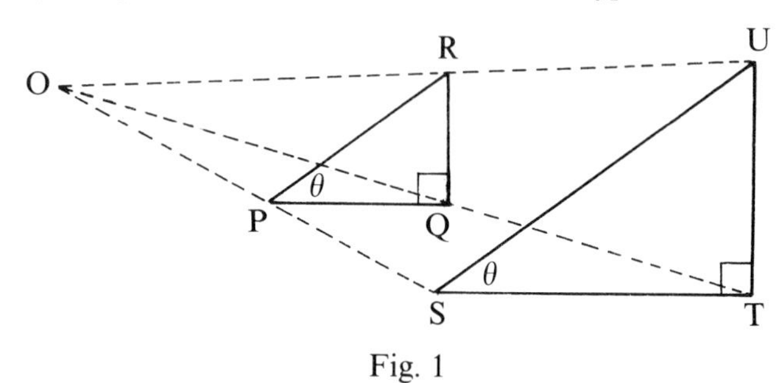

Fig. 1

In Fig. 1, $\triangle$STU is an enlargement of $\triangle$PQR with centre O. Suppose the scale factor is k. Then TU $= k \times$ QR and SU $= k \times$ PR and so

$$\frac{\text{TU}}{\text{SU}} = \frac{k \times \text{QR}}{k \times \text{PR}} = \frac{\text{QR}}{\text{PR}}$$

Thus for all right-angled triangles with an angle of size θ, the ratio

$$\frac{\text{side opposite } \theta}{\text{hypotenuse}} \text{ is the same. (Fig. 2)}$$

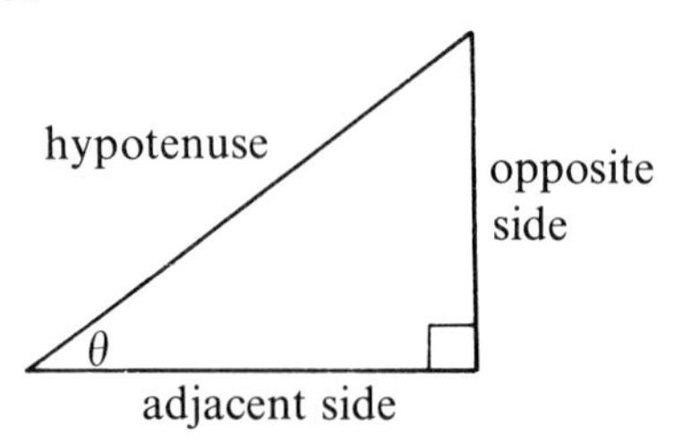

Fig. 2

This ratio is called the *sine* of θ and we write it as $\sin \theta$.

If you draw the triangle in Fig. 3 and measure BC you will find it is 3.5 cm approximately. Hence

$$\sin 26^\circ \simeq \frac{3.5}{8} = 0.44 \text{ to 2 d.p.}$$

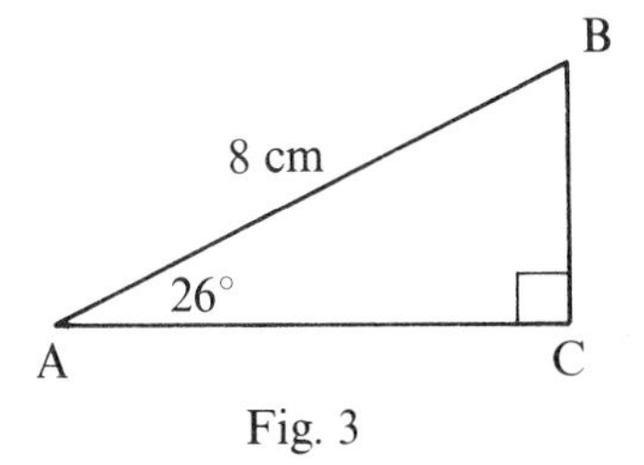

Fig. 3

Exercise 64

1. Draw right-angled triangles having $\theta = 40^\circ$ and hypotenuses of:
(i) 5 cm (ii) 6 cm (iii) 8 cm (iv) 10 cm.
In each case measure the opposite side and calculate a value for $\sin 40^\circ$, correct to 2 sig. fig.

2. Make a copy of Fig. 4 on graph paper using a radius of 10 cm and angles of 10° between the radii OA, OP_1, OP_2, etc.

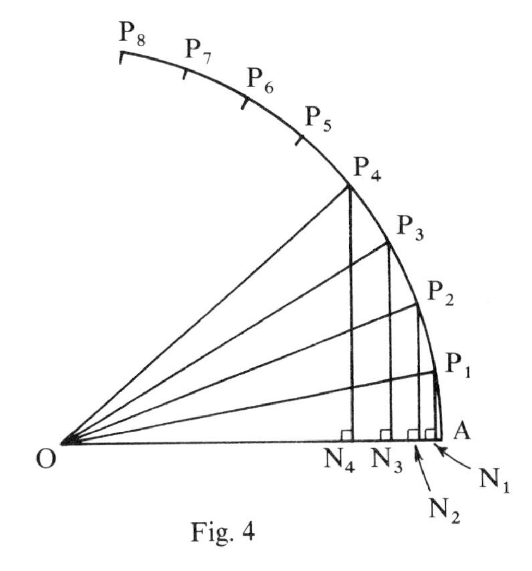

Fig. 4

Read off the heights P_1N_1, P_2N_2, P_3N_3, ... P_8N_8 and use them to find values for $\sin 10^\circ$, $\sin 20^\circ$, ... $\sin 80^\circ$, correct to 2 sig. fig.

For example, from $\triangle OP_2N_2$, $\sin 20^\circ = \dfrac{P_2N_2}{OP_2} = \dfrac{3.4}{10} = 0.34$

Copy and complete the table:

Angle	10°	20°	30°	40°	50°	60°	70°	80°
Sine		0.34						

From your table give an estimate of the value of sin 35° and of sin 55°

3. Using your results from Question **2**, draw a graph of the sines of angles from 0° to 90°. If your graph paper is large enough, use scales of 2 cm to 10° on the x axis and 2 cm to 0.1 on the y axis.

For 0° and 90° you need to think about the height of PN when $\widehat{POA}$ is: (a) very small (b) very near to 90°.

(i) From your graph check the estimate you made in Question **2** for sin 35° and sin 55°.

(ii) From your graph state the sines of:
15°, 23°, 38°, 56° and 76°.

(iii) From your graph state the angles which have sines of:
0.44, 0.56, 0.72, 0.90 and 0.95.

4. Use your table of Question **2** to calculate to 2 sig. fig. the opposite side when:

(i) $\theta = 40°$ and the hypotenuse is 20 cm.
(ii) $\theta = 60°$ and the hypotenuse is 6 cm.
(iii) $\theta = 70°$ and the hypotenuse is 8 cm.

5. Use your graph to calculate to 2 sig. fig. the opposite side when:

(i) $\theta = 35°$ and the hypotenuse is 40 cm.
(ii) $\theta = 45°$ and the hypotenuse is 9 cm.
(iii) $\theta = 64°$ and the hypotenuse is 4 cm.

6. Use your graph to find the value of θ if:

(i) the hypotenuse is 10 cm and the opposite side is 6 cm.
(ii) the hypotenuse is 20 cm and the opposite side is 15 cm.
(iii) the hypotenuse is 5 m and the opposite side is 2.3 m.

In Fig. 1, ST $= k \times$ PQ and SU $= k \times$ PR and so

$$\frac{\text{ST}}{\text{SU}} = \frac{k \times \text{PQ}}{k \times \text{PR}} = \frac{\text{PQ}}{\text{PR}}$$

Thus for all right-angled triangles with an angle of size θ, the ratio

$$\frac{\text{side adjacent to } \theta}{\text{hypotenuse}} \text{ is the same.}$$

This ratio is called the *cosine* of θ and we write it as $\cos \theta$.

Exercise 65

1. In the triangles you drew for Exercise **64**, Question **1**, measure the adjacent side in each case and calculate values for cos 40°.

2. Use your copy of Fig. 4 to obtain values for cos 10°, cos 20°, cos 30° . . . , cos 80° to 2 sig. fig. Show your results in a table. From the results estimate cos 35° and cos 55°.

3. Use your results of Question **2** to draw a graph of the cosines of the angles from 0° to 90°.
  (i) How does $\cos \theta$ change as θ increases? Does $\sin \theta$ change in the same way?
  (ii) From your graph check the estimates of cos 35° and cos 55° which you made in Question **2.**
  (iii) From your graph state the approximate values of cos 25°, cos 32°, cos 45°, cos 63° and cos 77°.
  (iv) From your graph state the angles which have cosines of 0.97, 0.79, 0.57, 0.38 and 0.12.

4. Calculate the adjacent side if:
  (i) $\theta = 25°$ and the hypotenuse is 20 cm,
  (ii) $\theta = 63°$ and the hypotenuse is 5 m.

5. Calculate θ if:
  (i) the hypotenuse is 20 cm and the adjacent side is 8 cm,
  (ii) the hypotenuse is 100 m and the adjacent side is 79 m.

In Chapter 11, we used tables of tangents. Tables of sines are used in the same way.

With tables of cosines you must be careful when using the difference columns on the right-hand side. Here is part of a cosine table.

NATURAL COSINES subtract

Degrees	0′	6′	12′	18′	24′	30′	36′	42′	48′	54′	Mean Differences				
		0.1°	0.2°	0.3°	0.4°	0.5°	0.6°	0.7°	0.8°	0.9°	1′	2′	3′	4′	5′
45	.7071						6997	6984				4			
46	.6947														

Notice that cos 45°36′ is 0.6997
and cos 45°42′ is 0.6984.

For a larger angle, we have a smaller cosine. On the 45° row there is 4 in the difference column headed 2′.

Hence cos 45°38′ is 0.6997 − 0.0004 (**not** + 0.0004)
that is, 0.6993.

Use your tables of sines and cosines to check your values in Exercise **64**, Question **2** and Exercise **65**, Question **2**.

EXAMPLE 1:

$$\frac{a}{8} = \sin 65° \simeq 0.9063$$

$$a \simeq 8 \times 0.9063 = 7.2504$$

BC is 7.25 cm, to 3 sig. fig.

$$\frac{b}{8} = \cos 65° \simeq 0.4226$$

$$b \simeq 8 \times 0.4226 = 3.3808$$

AC is 3.38 cm to 3 sig. fig.

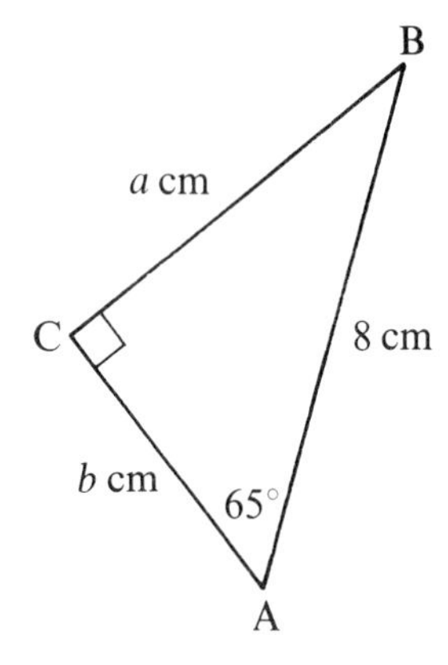

Fig. 5

Notice that the side opposite A has been marked a and the side opposite B has been marked b. This is the usual notation.

EXAMPLE 2:

$$\sin Q = \frac{4.3}{7} \simeq 0.6143$$

$$Q \simeq 37°54' \ (37.9°)$$

$$R \simeq 90° - 37°54'$$

$$= 52°6' \ (52.1°)$$

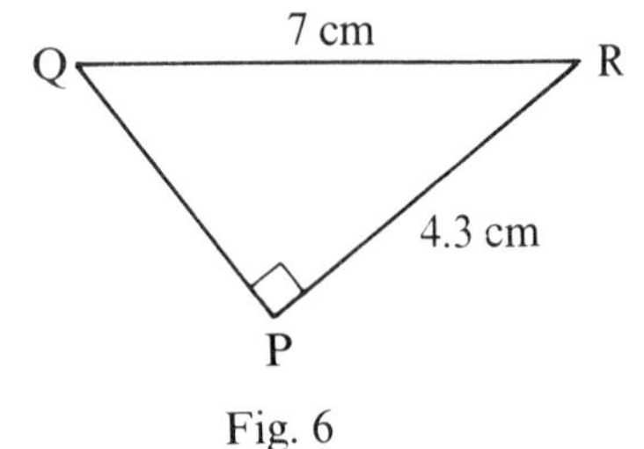

Fig. 6

Exercise 66

1. Use tables to write down the values of:

(i) sin 34.2° (ii) sin 76.6° (iii) sin 5.7°
(iv) sin 21°8′ (v) sin 35°45′ (vi) sin 69°52′.

2. Use tables to write down the values of:

(i) $\cos 20.3°$ (ii) $\cos 60°$ (iii) $\cos 83.6°$
(iv) $\cos 59°36'$ (v) $\cos 59°38'$ (vi) $\cos 23°16'$.

3. State the angles having sines of:

(i) 0.2924 (ii) 0.5678 (iii) 0.5680
(iv) 0.8545 (v) 0.8548 (vi) 0.7486

4. State the angles having cosines of:

(i) 0.2924 (ii) 0.5678 (iii) 0.5666
(iv) 0.8755 (v) 0.8751 (vi) 0.3703.

In Questions **5** to **10**, calculate the unknown sides, correct to 3 sig. fig:

5.

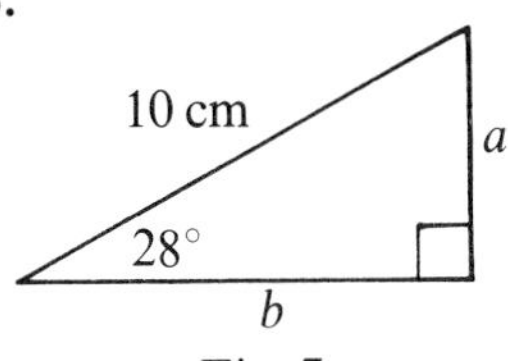

Fig. 7

6.

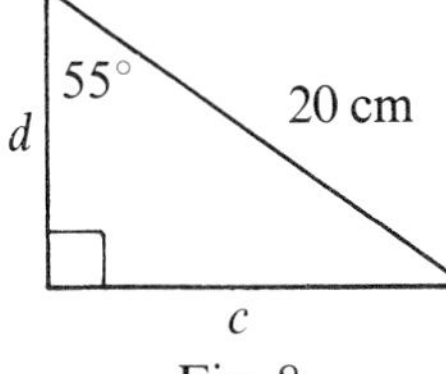

Fig. 8

7.

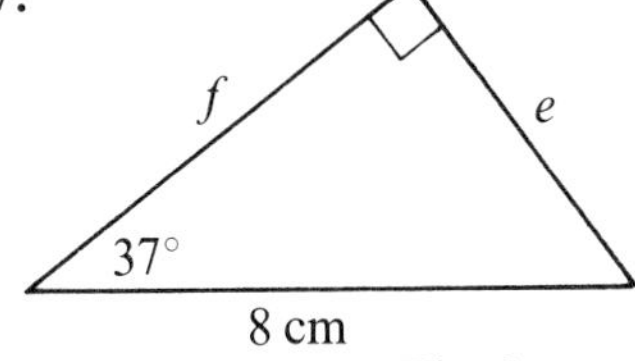

Fig. 9

8.

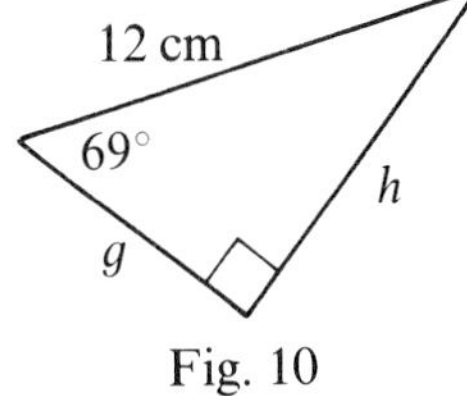

Fig. 10

9.

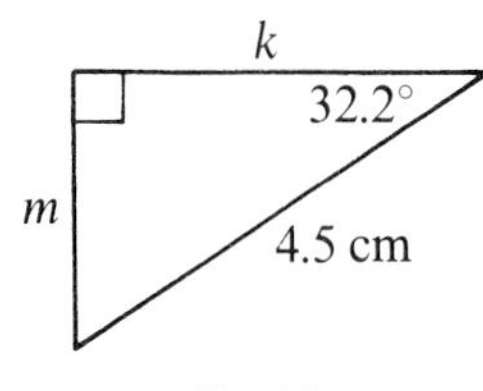

Fig. 11

10.

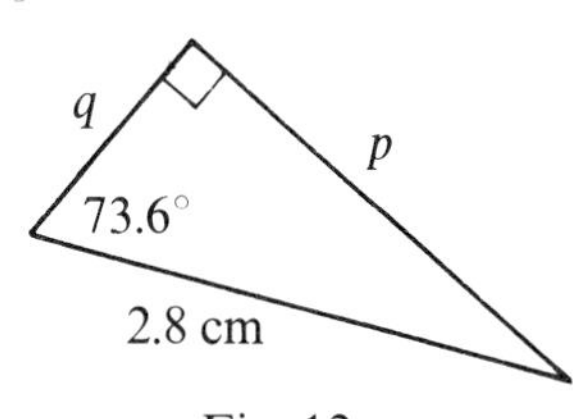

Fig. 12

In Questions **11** to **16**, calculate the size of the angle θ:

11.

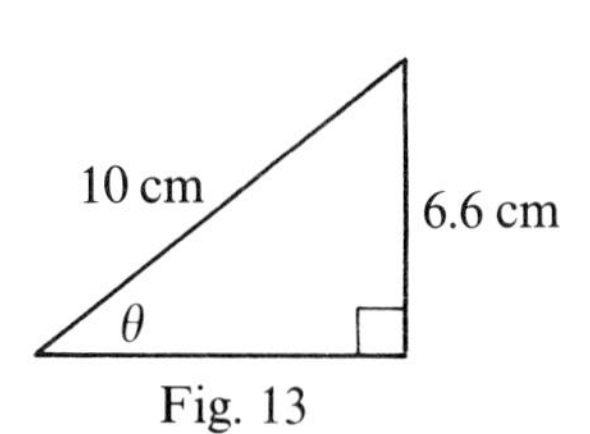

Fig. 13

12.

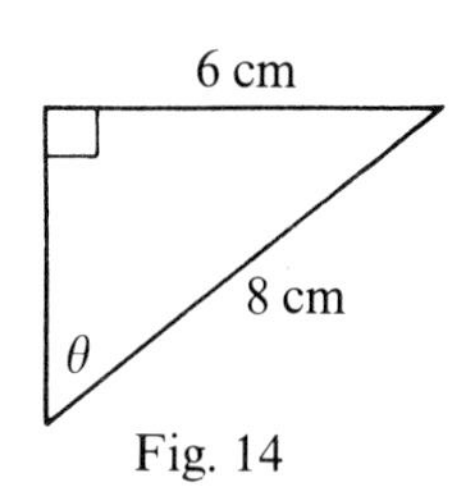

Fig. 14

13.

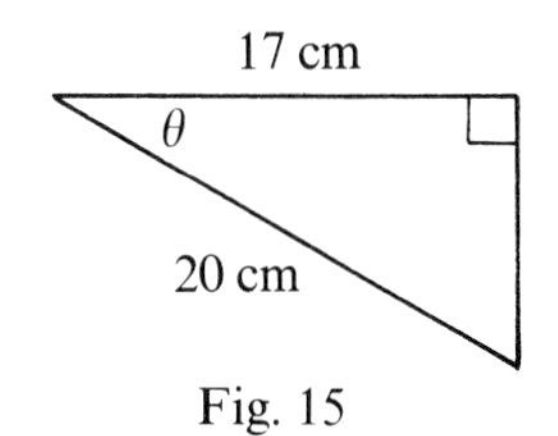

Fig. 15

14.

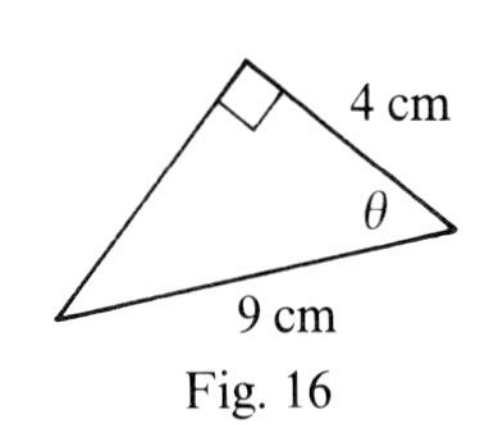

Fig. 16

15.

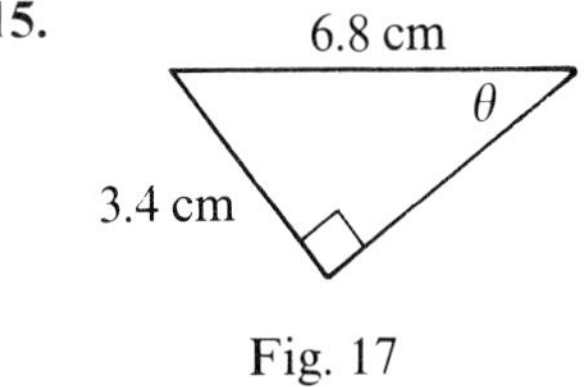

Fig. 17

16.

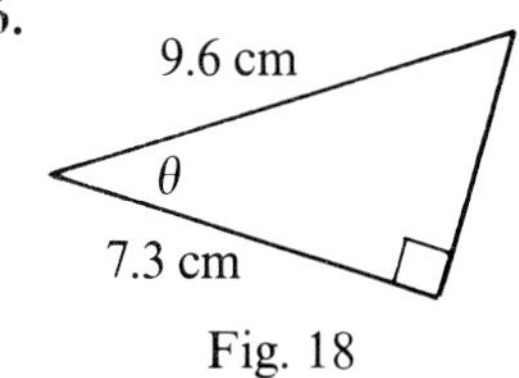

Fig. 18

17.

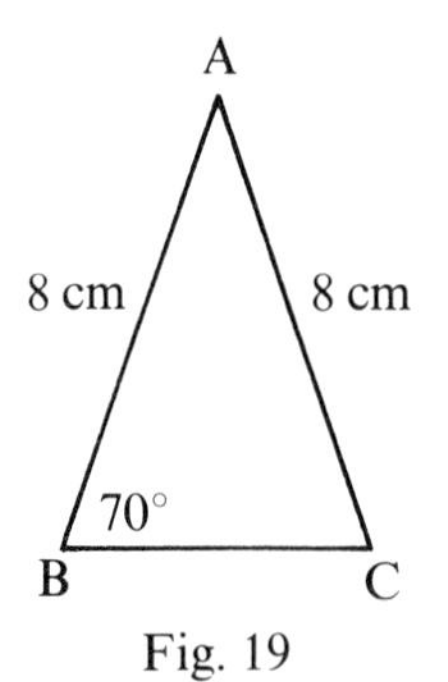

Fig. 19

Calculate BC (Draw altitude AN)

18.

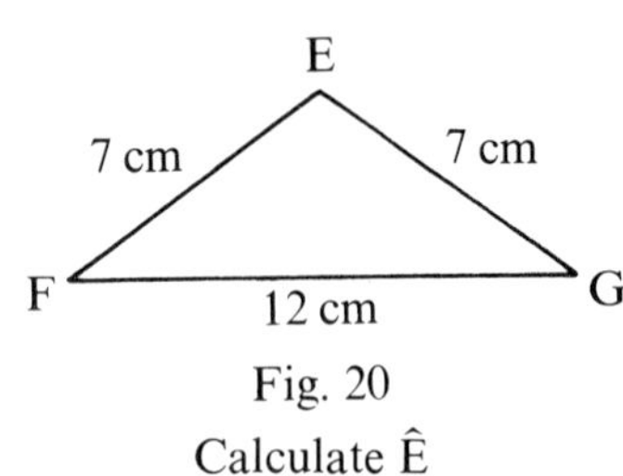

Fig. 20

Calculate Ê

19.

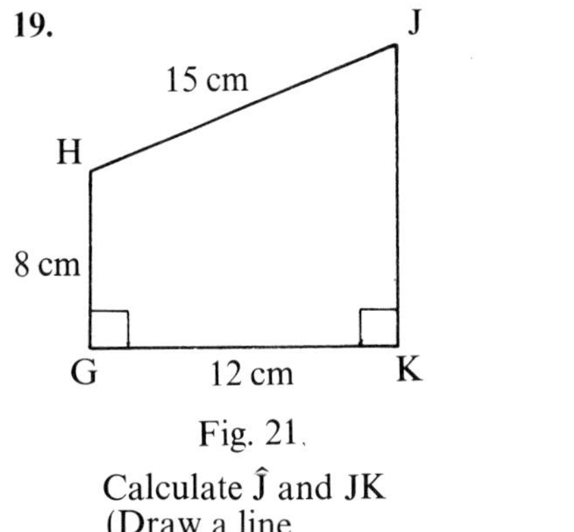

Fig. 21

Calculate $\hat{J}$ and JK
(Draw a line
from H)

20.

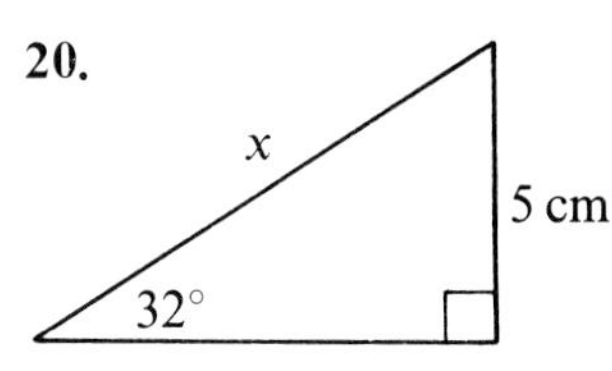

Fig. 22

Calculate x
(Use logarithms)

21.

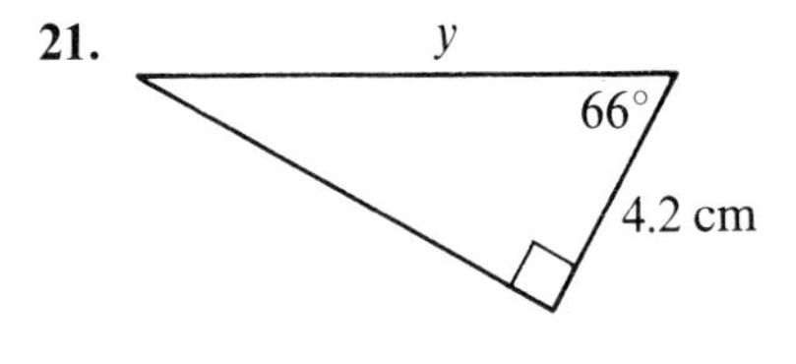

Fig. 23

Calculate y

22. (i) State the values of sin 40° and sin 80°.
$80° = 2 \times 40°$. Is $\sin 80° = 2 \times \sin 40°$? Give a reason for your answer using a diagram like Fig. 4.

(ii) $72° = 3 \times 24°$. Is $\cos 72° = 3 \times \cos 24°$ or $\frac{1}{3} \cos 24°$ or neither? Explain with the aid of a diagram.

23. From your tables write down the value of sin 38° and of cos 52°. Use Fig. 24 to explain why they are the same. Copy and complete:

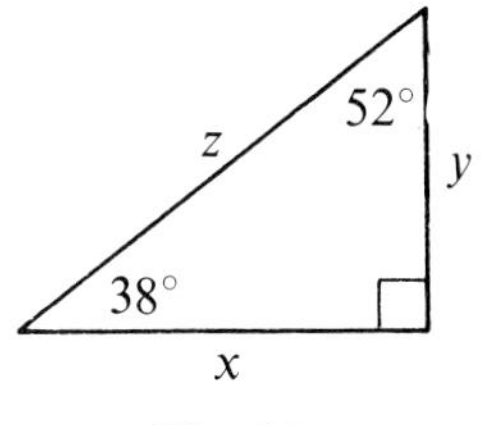

Fig. 24

(i) $\sin 22° = \cos \ldots$ (ii) $\cos 33.7° = \sin \ldots$

Exercise 67

1. A ladder of length 10 m is placed with its foot 3 m from a wall. What angle does it make with the horizontal?

2. A ladder of length 8 m makes an angle of 70° with the ground. How far up the wall does it reach?

3. A kite has a string of length 90 m which makes an angle of 62° with the ground. Calculate the height of the kite. (Assume that the string does not sag.)

4. A road rises 110 m in a distance of 1 km. Calculate the average angle between the road and the horizontal.

5. A mast is supported by three wire stays which are 30 m long and are fastened to the mast at a height of 23 m. Calculate the angle between a stay and the ground.

6. After take-off an aircraft climbs for 20 km at an angle of 25° to the horizontal. Calculate:

(i) the height it reaches (ii) the horizontal distance travelled.

7. An aircraft flies for 40 km on a bearing of 234°. How far:

(i) North or South
(ii) East or West is it of its starting point?

8. AB is a chord of a circle, centre O and radius 8 cm. If AB = 9 cm, calculate $\widehat{AOB}$.

9. A regular pentagon ABCDE has its vertices on a circle, centre O, radius 10 cm. Calculate:

(i) $\widehat{AOB}$
(ii) side AB
(iii) the shortest distance from O to AB
(iv) the area of the pentagon.

10. A rhombus has sides of 6 cm and two angles of 70°. Calculate:

(i) the length of the longer diagonal,
(ii) the length of the shorter diagonal,
(iii) the area of the rhombus.

REVISION PAPERS B

REVISION PAPER B1

1. (a) Simplify: $(4 - 2.3) \times 0.2$, 0.068×100 and $6.8 \div 100$.

(b) Express as fractions in their lowest terms:
$0.45, \quad 0.84, \quad 0.875, \quad 0.096.$

(c) Express as decimals: $\frac{7}{25}, \frac{13}{20}, \frac{47}{50}, \frac{19}{200}$.

2. (a) Solve $3(n - 4) = 2(n - 5)$.

(b) Solve $x - y = 4$, $7x + 2y = 1$.

3. The times taken for the six games of a certain chess match were: 49, 63, 10, 71, 53 and 66 minutes. Find the mean and median of the times. Comment on your results.

4. (a) Draw a diagram to illustrate the expansion of $(x + 5)(y + 4)$.

(b) *Without* drawing diagrams, obtain the expansions of:

(i) $(a + 3)(b + 5)$ (ii) $(c + 2)(c + 6)$
(iii) $(d - 8)(d + 3)$ (iv) $(5 - f)(7 - f)$.

5. (a) Triangle I has sides of 8 cm, 15 cm, and 17 cm;
Triangle II has sides of 9 cm, 16 cm and 25 cm;
Triangle III has sides of 12 cm, 16 cm and 20 cm.
Which triangles are right angled?

(b) In triangle PQR, angle $\hat{P} = 90°$, PQ = 7 cm and PR = 10 cm. Calculate QR correct to 2 significant figures.

6. A doctor gave the formula $S = 17 - \frac{1}{2}A$ for the number of hours of sleep needed by a person up to the age of 18 years. S is the number of hours of sleep and A is the age in years.

(i) How long is needed by a person aged: (a) 12 years (b) 18 years?
(ii) Make A the subject of the formula.
(iii) At what age does a person need: (a) 14 h (b) 10 h of sleep?

7. In triangle ABC, angle B = 90°.

(i) If angle A = 38° and AB = 6 cm, calculate BC.
(ii) If AB = 9 cm and BC = 7 cm, calculate angle C.

8. The table shows the distribution of teachers by age in a certain city in 1977.

Age in years	20–30	30–40	40–50	50–66
Number of teachers	150	400	250	200

The class 20–30 years contains those teachers aged 20 or more but less than 30 years. Draw a histogram using a suitable area scale. What is the modal class?

REVISION PAPER B2

1. (a) Write down the squares of: 30, 0.3 and 0.03.

(b) Write down, where possible, the exact square roots of: 400, 0.4, 0.04, 0.25 and 0.025.

(c) Given that $\sqrt{3.4} \simeq 1.84$ and $\sqrt{34} \simeq 5.83$, write down approximate values for:

$\sqrt{3400}$, $\sqrt{340}$, $\sqrt{0.34}$ and $\sqrt{0.034}$.

2. Factorise:

(i) $5a - 5b$ (ii) $c^2 - c$ (iii) $d^2 - 16$
(iv) $1 - f^2$ (v) $g^2 + 4g + 3$ (vi) $h^2 - 6h + 5$.

3. Work out the following and give the answers in standard form:

(i) $(0.07)^2$ (ii) $(0.03)^3$ (iii) 0.8×0.0055
(iv) $0.03 \div 5$ (v) $0.45 \div 0.0005$.

4. The number of engineering apprentices taken on each year by a certain firm from 1968 to 1977 is given as:

30, 26, 28, 35, 32, 31, 26, 25, 21, 16

Find the mean and median number of apprentices and comment on the data.

5. The diameter of a car wheel is 70 cm. Calculate its circumference, taking π as $3\frac{1}{7}$. How many revolutions does the wheel make when the car travels 1 kilometre?

6. Six girls shared equally a prize of £x. How much did each get? Later nine boys shared equally a prize of the same value. How much did each boy get?

If each boy received 50p less than each girl, set up an equation for x and solve it.

7. From the top of a cliff 80 m high, the angle of depression of a rowing boat is 32°. Calculate the distance of the boat from the cliff. A buoy is 200 m from the cliff. Calculate the angle of depression of the buoy from the top of the cliff.

8. On squared paper draw triangles ABC and PQR with vertices:

A(1, 3), B(4, 6), C(2, 7), P(3, 1), Q(6, 4), R(7, 2).

Show the line of symmetry as a broken line and state its equation. Mark as E the point (5, 7) and mark as F the reflection of E in the broken line. State the coordinates of F.

What would be the reflection of the point (17, 59) in the broken line?

REVISION PAPER B3

1. (a) Which of these are simple closed curves?

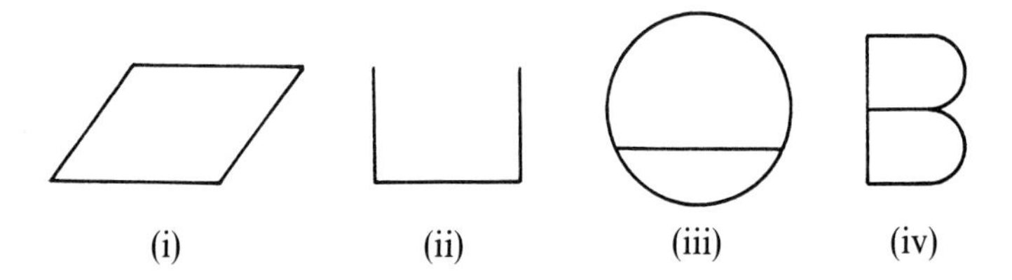

Fig. 1

(b) Which of these are topologically equivalent?

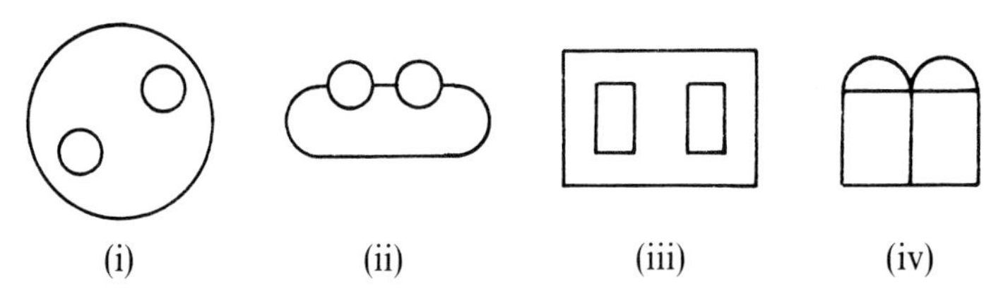

Fig. 2

2. (a) Give two examples where probability ideas are used without numbers.

(b) There is a probability of $\frac{3}{5}$ of getting a tail with a throw of a biased coin. What is the probability of getting a head with a throw of this coin?

3. (a) Expand:

(i) $(a - 7)(a + 7)$ (ii) $(b - 3)(2b - 1)$
(iii) $(5 - c)(3 + 2c)$ (iv) $(d - 4)^2$.

(b) Solve: $(x - 2)(x + 1) = (x - 3)(x + 4)$.

4. Use logarithms to calculate:

(i) the area of a circle of radius 8.7 cm, ($\pi = 3.14$)
(ii) the side of a cube which has a volume of 52 cm^3.

5.
$$\text{P} = \begin{pmatrix} 5 & -3 & 2 \\ 4 & 0 & -1 \end{pmatrix}, \quad \text{Q} = \begin{pmatrix} 2 & -1 \\ -2 & 5 \end{pmatrix},$$

$$\text{R} = \begin{pmatrix} 1 & 5 \\ 3 & 2 \\ -1 & 4 \end{pmatrix}, \quad \text{S} = \begin{pmatrix} 6 & 3 \\ 4 & 2 \end{pmatrix}$$

(i) Simplify where possible PQ, PR, RQ, QP and QS.
(ii) Solve the equation $4\text{X} = \text{QS}$.
(iii) Solve the equation $\text{Y} + \text{Q} = \text{PR}$.

6. A box contains a mixture of fivepenny and twopenny coins. There are 33 coins in the box and their value is £1.20. Form two equations and solve them to find the number of each type of coin.

7. (i) If $\theta = 35°$ and $b = 6.4$ cm, calculate p.
(ii) If $p = 9.7$ cm and $b = 8$ cm, calculate θ.

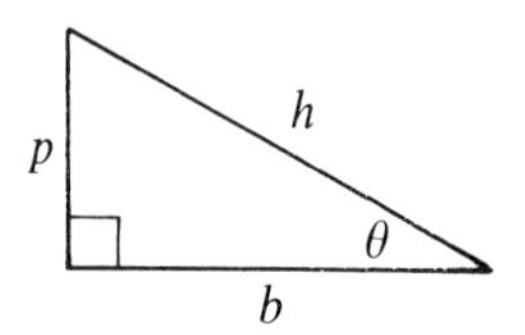

Fig. 3

8. A farmer wishes to fence off a rectangular piece of ground of area 400 m^2 using a wall on one side. (See Fig. 4.) Copy and complete the table.

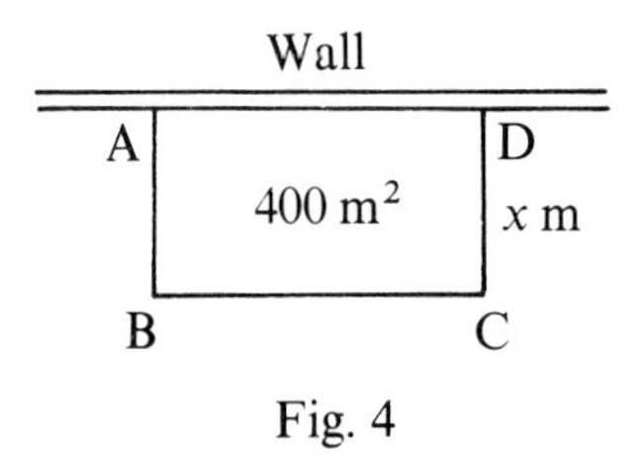

Fig. 4

AB (x metres)	5	10	15	20	25	30
BC	80		$26\frac{2}{3}$			
AB + BC + CD (ym)	90		$56\frac{2}{3}$			$73\frac{1}{3}$

Draw a graph to show the total length of fencing (y metres) for the values of x shown in the table.

From your graph state the value of x for which y is least.

Write down a formula for y in terms of x.

REVISION PAPER B4

1. Factorise, where possible:

(i) $a^2 + 9$ (ii) $b^2 - 9$ (iii) $c^2 + 8c - 9$
(iv) $d^2 + 10d - 9$ (v) $e^2 - 10e + 9$ (vi) $f^2 - 3f - 9$.

2. Rearrange each formula so that the stated letter is the subject:

(i) $r = 2n - 4$; n (ii) $I = \frac{1}{100}PR$; P
(iii) $y = 100 - x^2$; x (iv) $A = \pi r^2$; r
(v) $T = 2\sqrt{h}$; h.

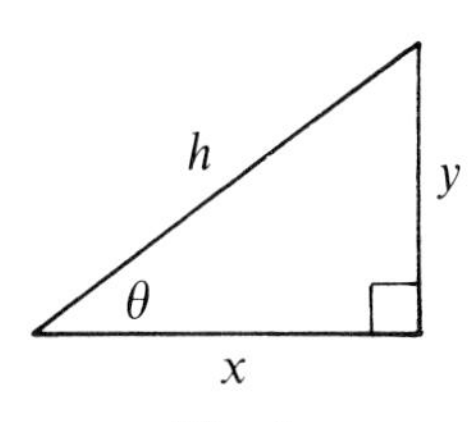

Fig. 5

3. (i) Using Fig. 5, copy and complete: $\tan\theta = \frac{\dots}{\dots}$ and $\sin\theta = \frac{\dots}{\dots}$

(ii) If $x = 5$ and $y = 8$, calculate θ.
(iii) If $h = 10$ and $y = 6$, calculate θ, x and the area of the triangle.

4. (a) Write down the cubes of -4, $\frac{3}{5}$, 0.2, a^4.
(b) Write down the cube roots of -27, $\frac{1}{8}$, 0.001, b^6.
(c) Use logarithms to calculate $(2.79)^3$ and $\sqrt[3]{2.79}$.

5. (a) Is it possible to have a regular polygon with an interior angle of:
(i) 108° (ii) 130° (iii) 150° (iv) 155° (v) 162°?
Where it is possible, state the number of sides.
(b) The exterior angles at three corners of a quadrilateral are 85°. Calculate the interior angle at the fourth corner.

6. Fig. 6 shows some members of a set of rectangles. Each has a long side on the x axis and a short side on the y axis. The dimensions of the rectangles are 9×3, 6×2, $4\frac{1}{2} \times 1\frac{1}{2}$, 3×1 units.

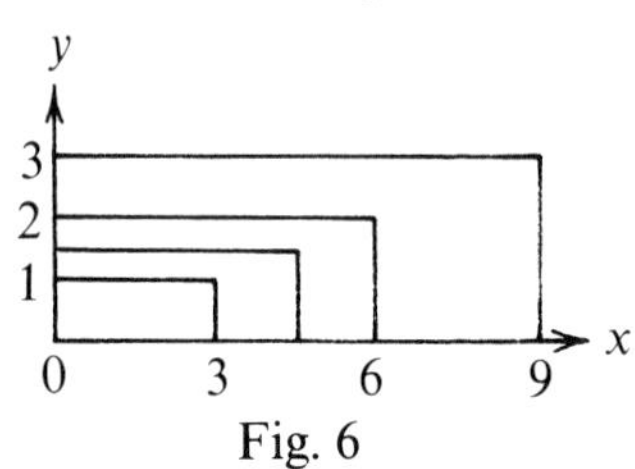

Fig. 6

What do you notice about the length and breadth of each rectangle? Does the rectangle measuring 5 m by $1\frac{2}{3}$ m have the same property?

Copy the figure on graph paper and draw a broken line through the corners not on the axes.

Which of the following is the equation of the line?
$x + y = 3$, $y = 3x$, $y = \frac{1}{3}x$, $xy = 3$
Starting from a point on the broken line, construct another member of the set of rectangles and state its dimensions.

7. There is a path of width 1.2 m round a circular pond of radius 4.8 m. Calculate, correct to 2 significant figures;
(i) the area of the pond, (ii) the area of the path.

8. The number of children in 80 families was as follows:

1	5	0	2	1	4	6	3	2	1	2	4	2	2	2	3	3	2	2	2
4	3	2	2	5	2	2	2	2	2	2	3	2	2	5	2	2	2	2	2
3	2	2	2	2	2	0	1	4	2	2	4	2	2	1	1	1	2	2	4
5	1	1	2	1	2	2	3	2	4	3	1	2	2	3	3	4	2	2	1

Prepare a frequency table and histogram. State the mode and median and calculate the mean. Comment on your results.

REVISION PAPER B5

1. (a) Write down the expansion of: $(a + b)(a - b)$
Use it to calculate the value of: (i) 42×38 (ii) 9.7×10.3.

(b) Write down the expansion of $(a + b)^2$ and use it to calculate, correct to 2 decimal places, 4.04^2 and 2.03^2.

2. (a) Simplify:

(i) $\frac{3ab}{6bc}$ (ii) $\frac{d^3}{d}$ (iii) $\frac{7(f - g)}{9(f - g)}$ (iv) $\frac{h^2 - 3h}{2h - 6}$.

(b) Express as a single fraction:

(i) $\frac{x}{3} - \frac{x}{4}$ (ii) $\frac{3}{y} + \frac{2}{y^2}$ (iii) $\frac{n}{15} \times \frac{10}{n}$ (iv) $\frac{a}{b} \div \frac{c}{d}$.

3. The first week's pay packets of 13 girls who left school together contained to the nearest pound,

17, 16, 18, 17, 27, 16, 19, 28, 18, 19, 17, 29, 19 pounds.

Find the mean and median values. Which do you think represents the data better?

4. Fig. 7 shows the front of a garage built against the wall of a house. Calculate the width AD (or BN) and the height of C above D.

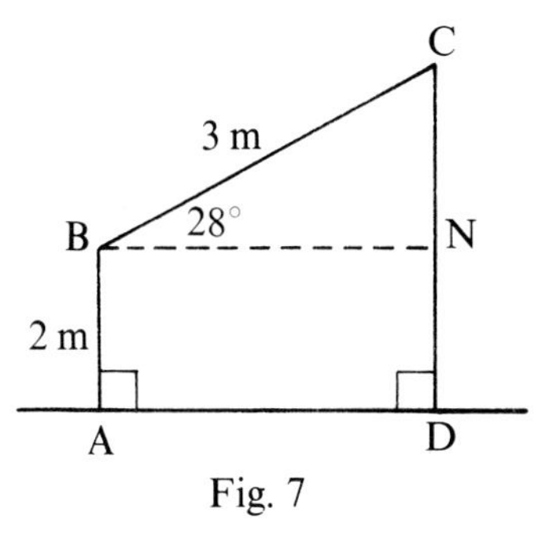

Fig. 7

5. (a) Find the value of:

(i) 30% of 50p (ii) 15% of £4
(iii) 44% of 1 kg (iv) 85% of 2 hours.

(b) Express the first amount as a percentage of the second:

(i) 90p, £6 (ii) £2.38, £7.

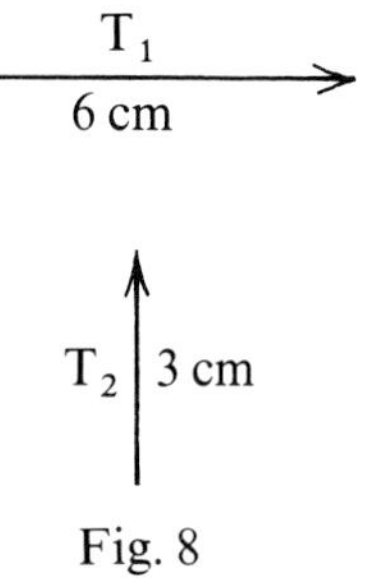

Fig. 8

6. Fig. 8 shows the translations T_1 and T_2. Draw diagrams to show the meaning of:

(i) $T_1 + T_2$ (ii) $T_1 - T_2$

7. Here are the results for four football teams:

A: won 4, drawn 5, lost 6; B: won 5, drawn 1, lost 3;
C: won 2, drawn 6, lost 3; D: won 5, drawn 2, lost 4.

Each team received 2 points for a win, 1 for a draw and 0 for a loss. Represent the data by two suitable matrices and multiply them together. What is the meaning of the numbers in your matrix?

Can you decide which is the best team?

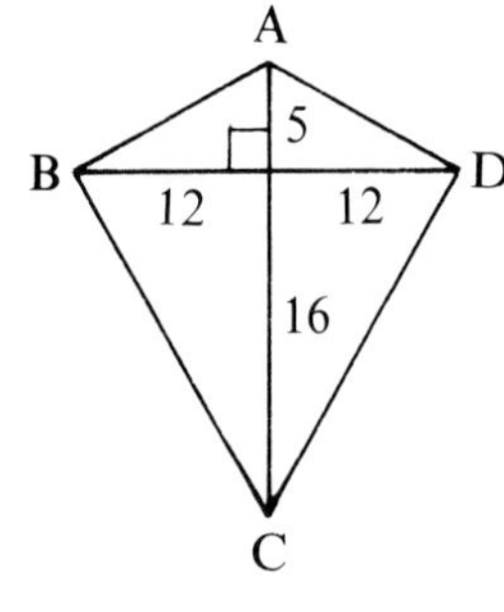

Fig. 9

8. Fig. 9 shows a kite, the dimensions being in centimetres. Calculate:

(i) AB
(ii) BC
(iii) the area of the kite.

17 · USING LOGARITHMS 2

The tables give the logarithms of numbers from 1 to 9.999. To find the logarithm of a number *greater than 10*, we express it in standard form.

The logarithm of 5.678 is 0.7542, from tables.

$$567.8 = 5.678 \times 100 = 10^{0.7542} \times 10^{2} = 10^{2.7542}$$

and so the logarithm of 567.8 is 2.7542.

We now consider the logarithms of numbers *less than 1.*

$$\begin{aligned} 0.05678 &= 5.678 \times 10^{-2} \\ &= 10^{0.7542} \times 10^{-2} = 10^{0.7542-2}, \end{aligned}$$

and so the logarithm of 0.05678 is $0.7542 - 2$.

We could combine the positive decimal and the negative integer to get -1.2458, but for using logarithms it is better not to do so. Instead we write it as $\bar{2}.7542$ which we read as 'bar two, point 7542'.

Thus for a number less than 1, the logarithm has a *negative* characteristic (whole number part) and a *positive* mantissa (decimal part).

Notice that $\bar{2}.7542$ is not the same as -2.7542. $\bar{2}.7542$ is $-2 + 0.7542$ whereas -2.7542 is $-2 - 0.7542$.

In Fig. 1, -2.7 and $\bar{2}.7$ are shown on a number line.

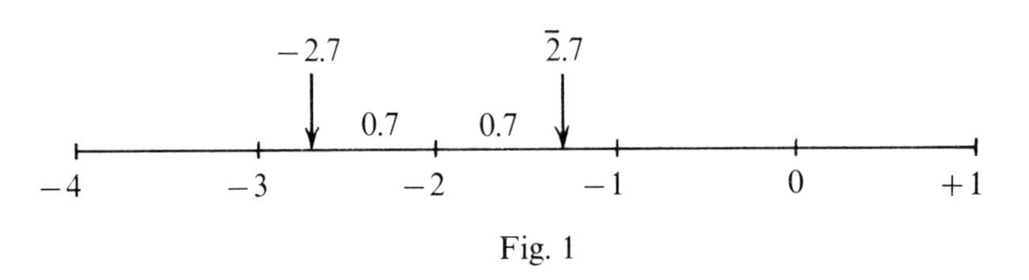

Fig. 1

Exercise 68

1. Copy and complete the following table:

Number	*Standard Form*	*Logarithm*
567.8	5.678×10^2	2.7542
56.78	... ...	...
5.678	5.678×10^0	0.7542
0.567 8	... ...	...
0.056 78	... ...	...
0.005 678	5.678×10^{-3}	$\overline{3}.7542$
0.000 567 8	... ...	...

2. Draw up a table as in Question **1** for the logarithms of 8500, 850, 85, 8.5, 0.85, 0.085.

3. Write each of the following in standard form and then state their logarithms:

(i) 0.258, 0.0258, 0.000 258.
(ii) 0.842, 0.000 842, 8420.
(iii) 0.0503, 0.503, 50.3.

4. (i) Draw a number line as in Fig. 1. Use arrows to show the positions of $\overline{3}.8$, $\overline{2}.2$ and $\overline{1}.6$.
(ii) Draw another number line and show the positions of -3.8, -2.2 and -1.6.

FINDING CHARACTERISTICS

The characteristic (whole number) part of a logarithm is equal to the index of 10 when the number is expressed in standard form.

Here is a method of obtaining the characteristic without writing the number in standard form:

Make sure that you have a nought before the decimal point and count the number of noughts up to the first significant figure.

0.000 216	4 noughts	$\overline{4}$
0.000 003 9	6 noughts	$\overline{6}$
0.0807	2 noughts	$\overline{2}$

ANTILOGARITHMS

Suppose that the logarithm of a number is $\bar{3}.4384$.

Looking up 0.4384 in the antilogarithm tables we obtain 2745.

As we have $\bar{3}$, we place 3 noughts in front and use the first nought in the unit place before the decimal point. This gives 0.002 745.

Exercise 69

1. State the characteristics (whole number parts) of the logarithms of:

(i) 0.0062 (ii) 0.0154 (iii) 0.803 (iv) 0.0009
(v) 52.03 (vi) 0.004 (vii) 0.2001 (viii) 3500.

Write down the logarithms of:

2. 0.725, 0.007 25, 72.5, 0.0725.
3. 0.0683, 6.83, 0.000 683, 6830.
4. 50.6, 0.506, 0.0506, 0.000 506.
5. 300, 0.03, 0.0003, 0.3.

Write down the numbers which have the following logarithms:

6. 0.265, $\bar{1}.265$, $\bar{2}.265$, $\bar{3}.265$.
7. 1.835, $\bar{1}.835$, $\bar{3}.835$, $\bar{5}.835$.
8. $\bar{2}.7184$, 0.7184, 2.7184, $\bar{1}.7184$.
9. $\bar{1}.093$, $\bar{3}.093$, 4.093, $\bar{4}.093$.

MULTIPLICATION

EXAMPLE:

82.4 × 0.0067

$\simeq$0.552 (3 sig. fig.)

(Rough check: 80 × 0.007 = 0.56)

Number		*Logarithm*
82.4	→	1.9159
×0.0067	→	$+\bar{3}.8261$
0.5521	←	$\bar{1}.7420$

Several different kinds of additions of logarithms can arise. Here are some examples:

$$\begin{array}{r} \bar{4}.3 \\ +1.4 \\ \hline \bar{3}.7 \\ \hline \end{array} \qquad \begin{array}{r} \bar{2}.5 \\ +\bar{3}.2 \\ \hline \bar{5}.7 \\ \hline \end{array} \qquad \begin{array}{r} 0.9 \\ +\bar{3}.4 \\ \hline \bar{2}.3 \\ \hline \end{array} \qquad \begin{array}{r} \bar{2}.8 \\ +\bar{4}.7 \\ \hline \bar{5}.5 \\ \hline \end{array} \qquad \begin{array}{r} \bar{2}.6 \\ +5.8 \\ \hline 4.4 \\ \hline \end{array}$$

Exercise 70

1. Write down the answers to:

(i) $\bar{2} + 5$ (ii) $\bar{6} + 1$ (iii) $2 + \bar{5}$ (iv) $6 + \bar{1}$ (v) $7 + \bar{7}$
(vi) $\bar{3} + \bar{4}$ (vii) $9 + \bar{2}$ (viii) $\bar{5} + 3$ (ix) $8 + \bar{2}$ (x) $4 + \overline{10}$.

2. Do the following additions:

(i) $\begin{array}{r} \bar{3}.5 \\ +4.2 \\ \hline \\ \hline \end{array}$ (ii) $\begin{array}{r} \bar{1}.6 \\ +\bar{2}.1 \\ \hline \\ \hline \end{array}$ (iii) $\begin{array}{r} 0.8 \\ +\bar{2}.5 \\ \hline \\ \hline \end{array}$ (iv) $\begin{array}{r} \bar{4}.6 \\ +0.7 \\ \hline \\ \hline \end{array}$ (v) $\begin{array}{r} \bar{1}.7 \\ +\bar{2}.7 \\ \hline \\ \hline \end{array}$

(vi) $\begin{array}{r} \bar{2}.8 \\ +3.8 \\ \hline \\ \hline \end{array}$ (vii) $\begin{array}{r} \bar{1}.4 \\ +0.9 \\ \hline \\ \hline \end{array}$ (viii) $\begin{array}{r} \bar{5}.7 \\ +1.6 \\ \hline \\ \hline \end{array}$ (ix) $\begin{array}{r} \bar{6}.3 \\ +\bar{2}.4 \\ \hline \\ \hline \end{array}$ (x) $\begin{array}{r} \bar{3}.9 \\ +\bar{2}.5. \\ \hline \\ \hline \end{array}$

Find the value, correct to 3 sig. fig., of:

3. 326×0.24
4. 13.9×0.62
5. 0.0794×2.37
6. 3.82×0.0564
7. 63.9×0.755
8. 0.084×0.946
9. 0.028×0.304
10. $4.635 \times 0.000\,762$
11. 0.633×92.67
12. $58.34 \times 0.008\,58$
13. 11.66×0.7033
14. 4.683×0.2592.

DIVISION

EXAMPLE:

$0.037 \div 0.684$

$\simeq 0.0541$ (3 sig. fig.)

(Rough check: $\dfrac{0.037}{0.7} = \dfrac{0.37}{7}$

$\simeq 0.05$)

Number	*Logarithm*
0.037 →	$\bar{2}.5682$
$\div$0.684 →	$-\bar{1}.8351$
0.05409 ←	$\bar{2}.7331$

Here are some examples of subtraction of logarithms. Study them carefully.

$$\begin{array}{r} \bar{1}.7 \\ -3.3 \\ \hline \bar{4}.4 \\ \hline \end{array} \quad \begin{array}{r} \bar{3}.7 \\ -\bar{1}.3 \\ \hline \bar{2}.4 \\ \hline \end{array} \quad \begin{array}{r} 3.7 \\ -\bar{1}.3 \\ \hline 4.4 \\ \hline \end{array} \quad \begin{array}{r} \bar{1}.3 \\ -3.7 \\ \hline \bar{5}.6 \\ \hline \end{array} \quad \begin{array}{r} \bar{3}.3 \\ -\bar{1}.7 \\ \hline \bar{3}.6 \\ \hline \end{array} \quad \begin{array}{r} 1.3 \\ -3.7 \\ \hline \bar{3}.6 \\ \hline \end{array}$$

Check the above by addition. For example, $\bar{4}.4 + 3.3 = \bar{1}.7$.

Exercise 71

1. Write down the answers to:

(i) $5 - 2$, $\bar{5} - 2$, $\bar{1} - 2$, $\bar{3} - 3$, $\bar{3} - 5$
(ii) $1 - \bar{3}$, $2 - \bar{1}$, $4 - \bar{5}$, $1 - \bar{1}$, $0 - \bar{6}$
(iii) $\bar{3} - 2$, $\bar{3} - \bar{2}$, $\bar{2} - \bar{3}$, $2 - \bar{3}$, $\bar{2} - 3$
(iv) $\bar{2} - \bar{1}$, $\bar{1} - \bar{2}$, $\bar{1} - 2$, $2 - \bar{1}$, $1 - 2$.

2. Do the following subtractions and check your answers by addition:

(i) $\begin{array}{r} \bar{3}.4 \\ -1.2 \\ \hline \\ \hline \end{array}$ (ii) $\begin{array}{r} \bar{4}.7 \\ -\bar{1}.3 \\ \hline \\ \hline \end{array}$ (iii) $\begin{array}{r} \bar{5}.6 \\ -\bar{7}.4 \\ \hline \\ \hline \end{array}$ (iv) $\begin{array}{r} 0.9 \\ -\bar{3}.2 \\ \hline \\ \hline \end{array}$ (v) $\begin{array}{r} 2.6 \\ -\bar{3}.1 \\ \hline \\ \hline \end{array}$

(vi) $\begin{array}{r} 1.4 \\ -3.8 \\ \hline \\ \hline \end{array}$ (vii) $\begin{array}{r} \bar{2}.3 \\ -\bar{1}.6 \\ \hline \\ \hline \end{array}$ (viii) $\begin{array}{r} 0.7 \\ -\bar{2}.9 \\ \hline \\ \hline \end{array}$ (ix) $\begin{array}{r} \bar{1}.5 \\ -5.7 \\ \hline \\ \hline \end{array}$ (x) $\begin{array}{r} 2.6 \\ -\bar{4}.9 \\ \hline \\ \hline \end{array}$

3. Do the following subtractions and check your answers by addition:

(i) $\begin{array}{r} 0.8 \\ -2.4 \\ \hline \\ \hline \end{array}$ (ii) $\begin{array}{r} \bar{2}.9 \\ -\bar{3}.2 \\ \hline \\ \hline \end{array}$ (iii) $\begin{array}{r} \bar{2}.7 \\ -\bar{1}.9 \\ \hline \\ \hline \end{array}$ (iv) $\begin{array}{r} \bar{3}.1 \\ -\bar{5}.4 \\ \hline \\ \hline \end{array}$ (v) $\begin{array}{r} \bar{2}.7 \\ -4.5 \\ \hline \\ \hline \end{array}$

(vi) $\begin{array}{r} 3.6 \\ -\bar{5}.9 \\ \hline \\ \hline \end{array}$ (vii) $\begin{array}{r} \bar{6}.6 \\ -2.9 \\ \hline \\ \hline \end{array}$ (viii) $\begin{array}{r} 4.3 \\ -\bar{4}.7 \\ \hline \\ \hline \end{array}$ (ix) $\begin{array}{r} \bar{1}.0 \\ -\bar{3}.6 \\ \hline \\ \hline \end{array}$ (x) $\begin{array}{r} \bar{4}.5 \\ -6.7. \\ \hline \\ \hline \end{array}$

Evaluate, correct to 3 sig. fig:

4. $9.46 \div 72.8$
5. $5.28 \div 9.27$
6. $0.0798 \div 0.237$
7. $0.392 \div 56.4$
8. $6.39 \div 0.0755$
9. $0.838 \div 0.009\,46$
10. $0.028 \div 0.003\,68$
11. $4.72 \div 0.0766$
12. $0.0632 \div 0.9263$
13. $0.583 \div 8.724$
14. $0.0708 \div 42.65$
15. $0.016\,32 \div 0.0761$

POWERS

EXAMPLE:

$$(0.388)^3$$
$$= 0.0584$$
(3 sig. fig.)

(Rough check:
$(0.4)^3 = 0.064$)

Number	*Logarithm*
$0.388 \rightarrow$	$\bar{1}.5888$
$(0.388)^3 \rightarrow$	$\bar{1}.5888 \times 3$
	$= (\bar{1} + 0.5888) \times 3$
	$= \bar{3} + 1.7664 =$
$0.058\,39 \leftarrow$	$\bar{2}.7664$

Exercise 72

1. Write down the answers to:

(i) $\bar{1}.3 \times 2$ (ii) $\bar{3}.6 \times 2$ (iii) $\bar{4}.7 \times 3$ (iv) $\bar{1}.9 \times 4$
(v) $\bar{3}.2 \times 6$ (vi) $\bar{1}.8 \times 5$ (vii) $\bar{1}.24 \times 3$ (viii) $\bar{4}.67 \times 4$.

Evaluate, correct to 3 sig. fig.

2. $(0.246)^3$ **3.** $(0.929)^3$ **4.** $(0.716)^5$ **5.** $(0.804)^4$
6. $(25.67)^3$ **7.** $(0.2567)^3$ **8.** $(8.375)^4$ **9.** $(0.8377)^4$
10. $(6.84 \times 0.032)^2$ **11.** $(5.25 \div 7.04)^3$.

Give the answers to the following in standard form:

12. $(0.0365)^4$ **13.** $(0.007\,42)^3$ **14.** $(0.138)^8$ **15.** $(0.05)^{10}$.

ROOTS

$$\sqrt[3]{0.0585} = \sqrt[3]{\frac{58.5}{1000}} = \frac{\sqrt[3]{58.5}}{\sqrt[3]{1000}}$$
$$\simeq \frac{3.883}{10} = 0.3883$$
$$\simeq 0.388 \text{ (3 sig. fig.)}$$

Number	*Logarithm*
$58.5 \rightarrow$	1.7672
$\sqrt[3]{58.5} \rightarrow$	0.5891
$3.883 \leftarrow$	

We introduce 1000 because its cube root is 10.

For $\sqrt[4]{0.0585}$ we write $\sqrt[4]{\dfrac{585}{10\,000}}$ because the fourth root of 10000 is 10.

Exercise 73

1. Copy and complete:

(i) $0.716 = \dfrac{}{1000}$ (ii) $0.0716 = \dfrac{}{1000}$

(iii) $0.529 = \dfrac{}{10000}$ (iv) $0.0148 = \dfrac{}{100\,000}$.

Evaluate, Questions **2** to **12**, correct to 3 sig. fig.

2. $\sqrt[3]{0.716}$ **3.** $\sqrt[3]{0.0716}$ **4.** $\sqrt[4]{0.529}$ **5.** $\sqrt[4]{0.0529}$

6. $\sqrt[3]{0.004\,67}$ **7.** $\sqrt[4]{0.902}$ **8.** $\sqrt[5]{0.0148}$ **9.** $\sqrt[5]{0.004\,08}$

10. $\sqrt[3]{0.000\,25}$ **11.** $\sqrt[4]{0.3684}$ **12.** $\sqrt[4]{0.071\,85}$

Exercise 74

Evaluate, correct to 3 sig. fig.

1. 739×0.052 **2.** $0.739 \div 52$

3. $(0.648)^3$ **4.** $\sqrt[3]{0.648}$

5. 0.1042×0.318 **6.** $0.0104 \div 0.000\,318\,4$

7. $\sqrt[4]{0.8766}$ **8.** $(0.8766)^4$

9. $\sqrt[3]{(0.29 \times 0.84)}$ **10.** $(0.43 \div 0.96)^3$.

11. Calculate the area of a square having a side of 0.837 metres.

12. Calculate the length of the edge of a cube which has a volume of 0.78 m^3.

13. A book has 278 pages and a thickness of 20.8 mm. Calculate the thickness of each sheet.

14. A new machine costing £3270 is placed in a factory. At the end of each year its value is only 0.8 of its value at the beginning of that year. Explain why its value after 10 years is £3270 $\times$ $(0.8)^{10}$. Calculate this value.

15. Use logarithms to calculate $(0.6)^{10}$ and compare your answer with the exact value which is 0.006 046 617 6.

18 · FLOW DIAGRAMS

SIMPLE DIAGRAMS

Suppose that you wish to send a letter to a friend. You must do the following: address an envelope, put the envelope in a letter box, write the letter, seal the envelope, stick a stamp on the envelope, place the letter in the envelope. Of course, this is not the correct order for the actions. A suitable order is shown in Fig. 1. Is this the only possible order? Is it the best order?

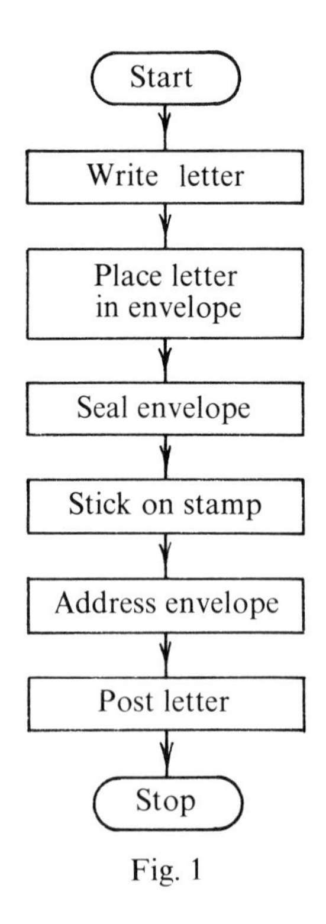

Fig. 1

Fig. 1 is an example of a flow diagram. Flow diagrams are useful for organising the way in which a job is to be done.

Notice the following features of the diagram:

(i) the shape of the termination boxes containing ‘start’ and ‘stop’.
(ii) the shape of the instruction boxes,
(iii) the arrows linking the boxes.

Exercise 75

1. Arrange each of the following as a flow diagram. Remember to use ‘start’ and ‘stop’.

(a) [Open door] [Place key in lock] [Close door] [Enter]
[Turn key] [Take key out of lock]

(b) [Go to school] [Wash] [Get up] [Have breakfast] [Dress]

(c) [Get out of train] [Get into train] [Give up ticket]
[Go to station] [Buy ticket]

(d) [Climb out] [Change into clothes] [Enjoy swim]
[Dive into pool] [Change into trunks or costume]
[Dry with towel]

2. Given the numbers and symbols

6 9 54 $\div$ $=$

we can use them for the statement $54 \div 6 = 9$. Is any other statement possible?

Arrange the following to make correct statements:

(a) $= \quad \times \quad 12 \quad {}_{4}84 \quad 7$
(b) $2 \quad 16 \quad =$
(c) $(a + b) \quad b^2 \quad a^2 \quad (a - b) \quad = \quad -$
(d) $\{p, q\} \quad \{q\} \quad \{q, r\} \quad = \quad \cap$

3. Write flow diagrams for three of the following:

(a) Making a pot of tea
(b) Putting on a record
(c) Washing your hands
(d) Obtaining tea from a vending machine
(e) Taking your dog for a walk.

4. (a) Fig. 2 is a flow diagram for finding the value of $3(x + 2)^2$ for a given value of x. If $x = 5$ we get 5, 7, 49, and finally 147. Work through the diagram with $x = 4$ and with $x = 8$.

(b) Draw a flow diagram to find the value of $(2x + 1)^2$ for a given value of x. Use it with $x = 3$ and with $x = 4$.
(c) Draw a flow diagram to find the value of $(2x - 7) \div 5$ for a given value of x. Use it with $x = 11$ and $x = 8$.

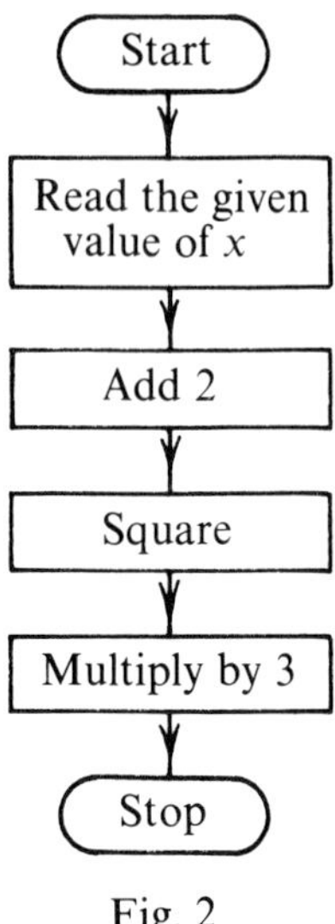

Fig. 2

5. Make up a series of actions like those in Question **1**. Mix them up and hand them to a neighbour to use for a flow diagram. Check his diagram.

6. Make up a simple mathematical statement like those in Question **2**. Mix up the symbols and numbers. Hand them to a neighbour and ask him to form the statement.

7. The formula for the area of a circle is $A = \pi r^2$. Use the following for a flow diagram: Divide by 2. Multiply by π. Square the number. Measure the diameter. (Test your diagram taking a diameter of 20 cm and 3.14 for π.)

8. Draw flow diagrams for the following:

(a) You are given two angles of a triangle and wish to calculate the third. Test your diagram using 70° and 65°.
(b) You are given two sides of a right-angled triangle and wish to calculate the hypotenuse. Test your diagram using 12 and 5.
(c) Finding the square root of a three figure number using tables.

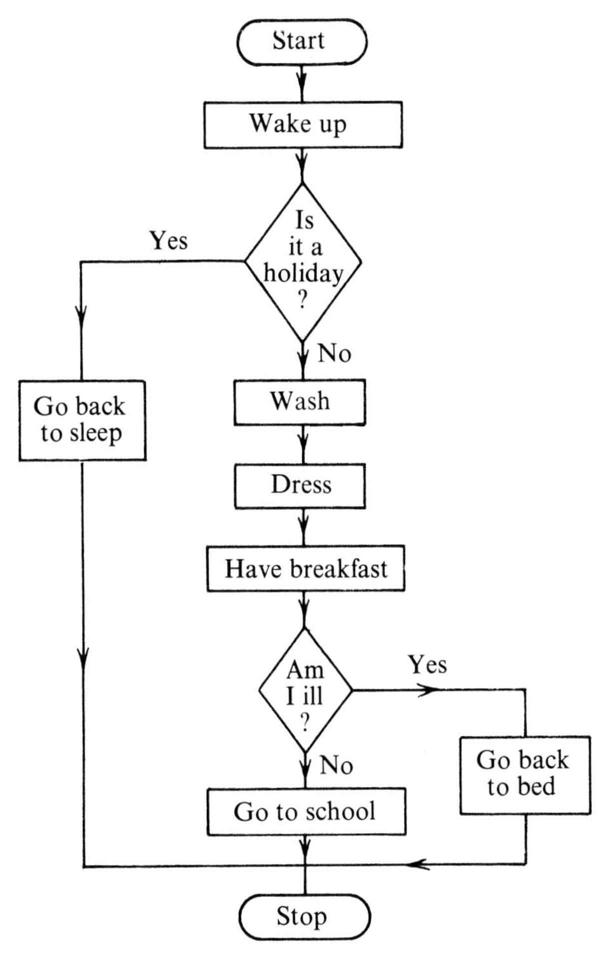

Fig. 3

USING DECISION BOXES

In Figs. 3 and 4, we have a new type of box—a decision box, in which a question is placed. Notice the shape. The question must be worded so that the answer is either 'Yes' or 'No'.

Fig. 3 does not need any explanation. Fig. 4 is for a shop assistant at sale time. All articles marked more than £10 are reduced by 20%: the others are reduced by 10%. Use the diagram for articles which are marked £8, £16 and £10.

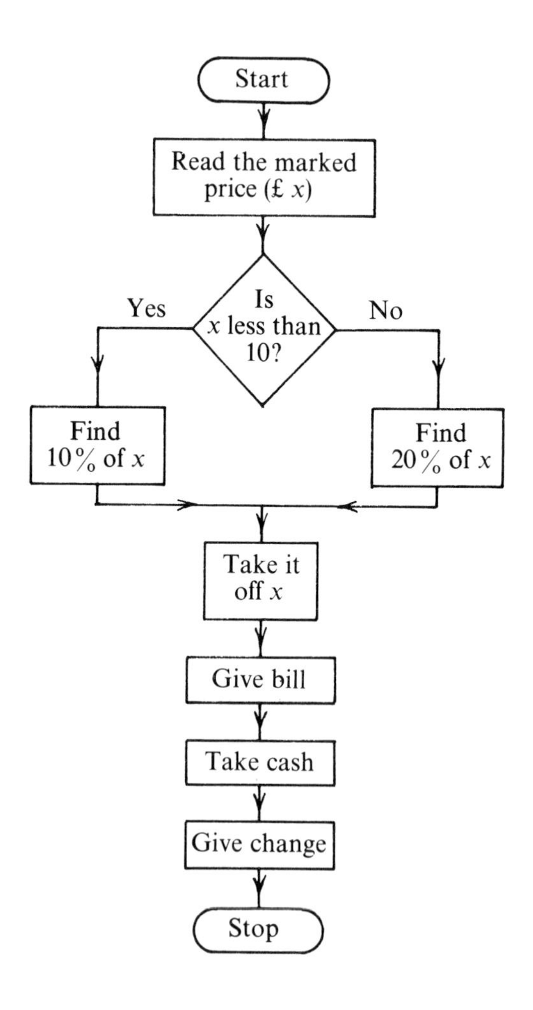

Fig. 4

Exercise 76

1. Which of the following questions are suitable for decision boxes? It must be possible to answer them with 'Yes' or 'No'. When a question is not suitable, try to rephrase it.
   - (i) Is the water warm enough?
   - (ii) Is 5 a factor of the number?
   - (iii) Is x greater or less than y?
   - (iv) Which way should I turn at the T junction?
   - (v) Can you change a £1 note?
   - (vi) Did Tom or Joan win the point?
   - (vii) Did the coin show a head or a tail?
   - (viii) What do you think of football?

2. A salesman receives 3% commission in a week in which his sales are less than £500 in value and 5% in a week in which they are £500 or more. Form the following into a flow diagram:

   Calculate 3% of £x. Calculate 5% of £x. Write down the sales figure (£x).

   Is x less than 500? Start. Stop.

   Test your diagram for sales of £440 and £790.

3. In the country of Percentia there is a sales tax of 5% on every article marked less than 50 percs and 10% on every article marked 50 percs or more.

   Draw a flow diagram similar to Fig. 4.

   Use your diagram to find how much must be paid for articles marked 40, 60, 170 and 50 percs.

4. Assemble the following into a flow diagram for finding whether or not a given number has 20 as a factor:

   Is the second digit from the right even? Does the number end in nought? Read the number. Write '20 is a factor'. Write '20 is not a factor'. Stop. Start.

   Test your diagram using the numbers 7365, 840 and 970.

5. A man is due at a factory at 8.30. A bus stops outside his house at 8.20 and gets him to work on time, but he likes to walk to work if he is ready by 8.00 and it is fine. Draw a flow diagram using the questions: 'Is it after 8.00? Is it fine? Is it after 8.20?'

6. Write a flow diagram for boiling an egg. For the decision box use 'Do I want it hard or soft?'

7. Draw a flow diagram for choosing the TV channel you want.

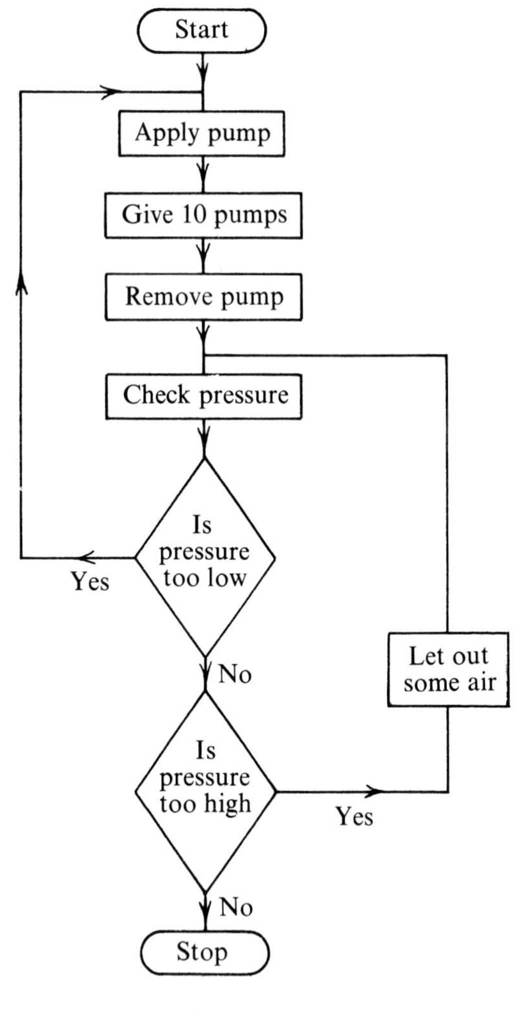

Fig. 5

LOOPS

Fig. 5 shows what a motorist might do if he finds that the pressure of one of his tyres is low. This diagram shows the use of loops for repeating instructions.

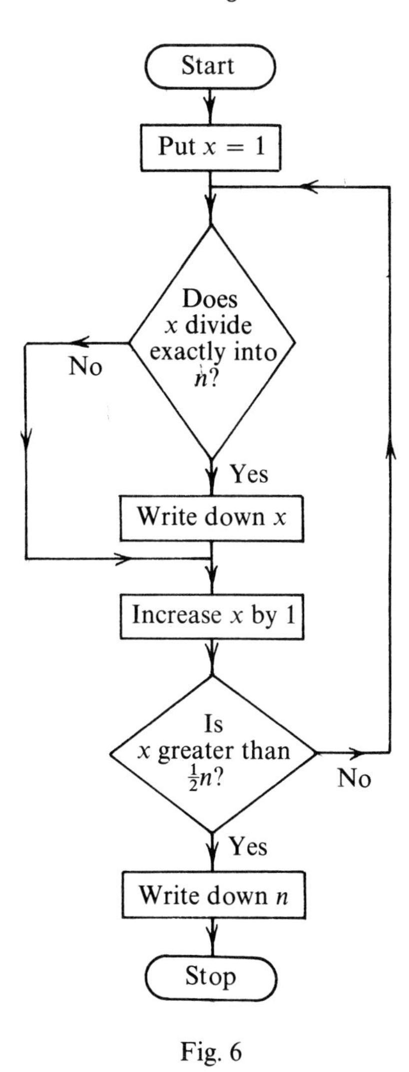

Fig. 6

Fig. 6 shows a diagram for obtaining the factors of a number which is called n. If we start with 12 as the value of n, then x takes the values 1, 2, 3, 4, 5, 6. For 1, 2, 3, 4 and 6 we write down the numbers. When $x = 5$ we bypass the instruction 'Write down x.' When

$x = 7$ the answer to 'Is x greater than $\frac{1}{2}n$?' is 'yes' and so we write down 12 and stop.

Work through the diagram with $n = 15$ and $n = 20$.

Exercise 77

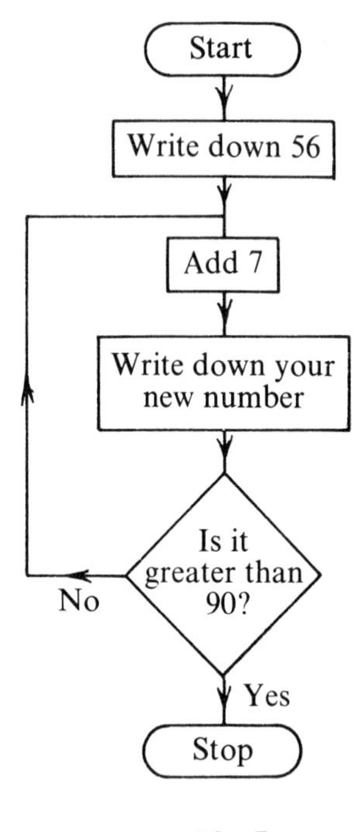

Fig. 7

1. Work through the diagram in Fig. 7 and explain what it does.
2. Work through the diagram in Fig. 8 setting out the results in a table like this:

x	1	3	5	...
Sum	0	1	4	

   Explain what the diagram does.
3. Draw a flow diagram to find the sum of the numbers 1, 2, 3, 4, 5, up to 10. Work through it. You should get the answer 55.

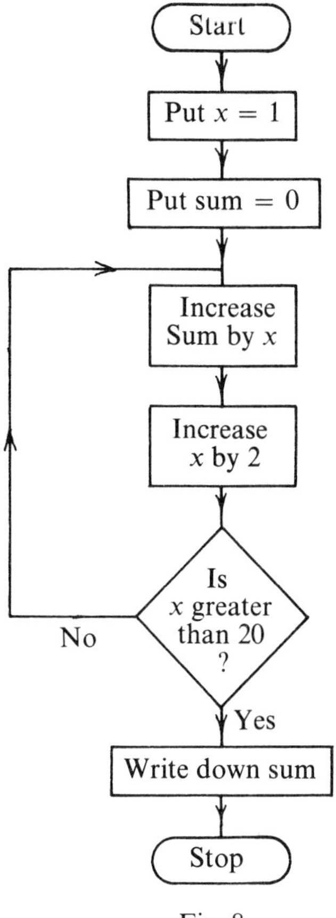

Fig. 8

4. Draw a flow diagram to find the sum of the numbers 1^2, 2^2, 3^2, up to 10^2. Work through it setting out your working as in Question **2.**

5. Draw a flow diagram to list the powers of 3 (3, 9, 27, 81, etc) until a number larger than 2000 is obtained.

Draw flow diagrams for the following:

6. Adjusting the temperature of a bath.

7. Crossing the road.

8. Making a telephone call.

9. Either: (i) planing a piece of wood to a required thickness.
or (ii) boiling potatoes correctly.

10. Mixing blue and yellow paint to get a shade of green which is needed.

11. In a certain game between two players, A and B, the score has reached 20–20. The play must continue until one player is 2 points ahead of the other.

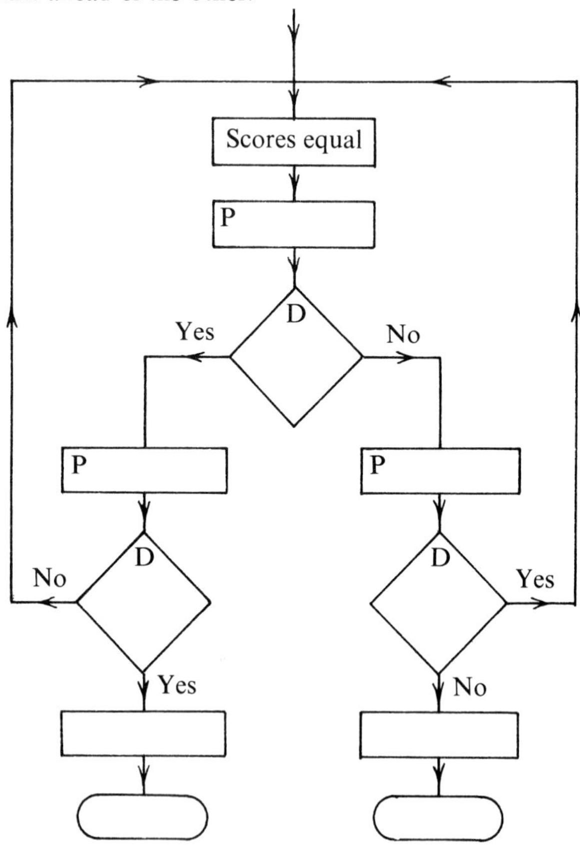

Fig. 9

Copy Fig. 9 and fill in the boxes so that you have a flow diagram for the rest of the game. In each box marked P write 'Play a point'. In each decision box (D) use the question 'Did A win the point?' Use also 'Game to A' and 'Game to B'.

19 · VECTORS

'To go from London to Bristol you have to travel 180 kilometres.' This is true, but it does not give you enough information to get to Bristol. It is necessary to add 'due West'. Otherwise you might arrive at Birmingham or Norwich or some other place which is 180 km from London.

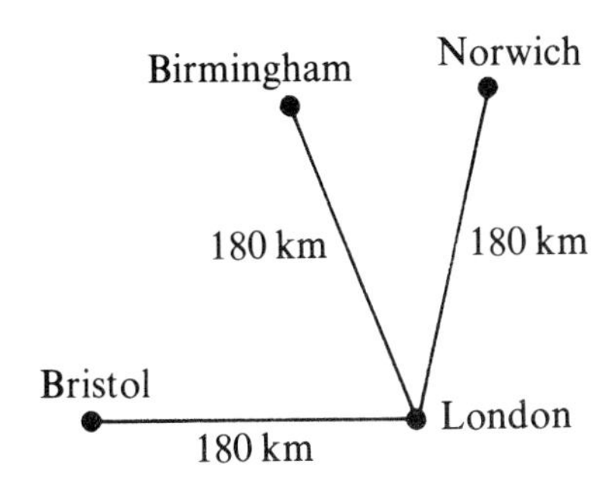

Fig. 1

To describe a movement (or displacement) both the size and direction must be stated. Quantities which have direction as well as size are called *vector quantities* or just *vectors*. Two other examples are velocity and force. An aircraft can have a velocity of 600 km per hour in the direction 075°

Some quantities do not have direction. They only have size. There are some examples in these sentences: 'Jim is 16 years old.' 'This record cost £2.99.' 'The journey took 5 hours.' Such quantities are called *scalar quantities* or just *scalars*.

We can represent a vector quantity by an arrow. The arrow is drawn in the correct direction and its length represents the size of the vector on a suitable scale. Some examples are shown in Fig. 2.

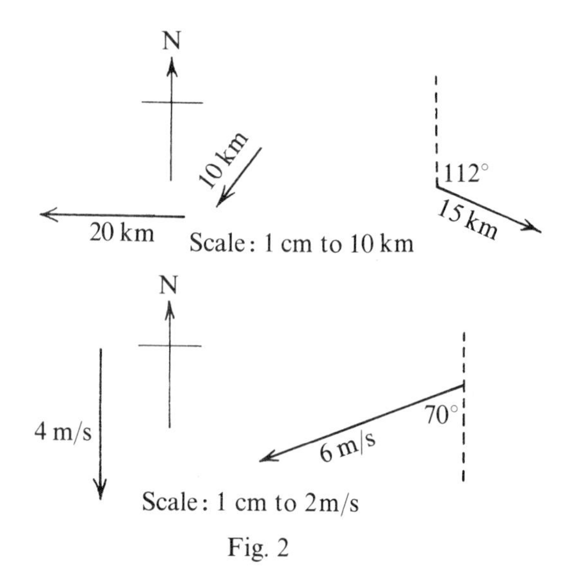

Fig. 2

Exercise 78

1. For each sentence state whether the quantity is a scalar or a vector quantity.

(a) There are 17 books on the shelf.
(b) The boat travelled at 10 knots in a northerly direction.
(c) The area of the hall is 150 m^2.
(d) The city houses a million people.
(e) The ball was kicked towards the goal at 30 m per second.
(f) It is a 150 watt bulb.

2. Which of the following situations involve vector quantities?

(a) Paying £1 for a pair of socks.
(b) Taking a throw-in at soccer.
(c) Passing a rugby ball.
(d) Taking part in a tug-of-war.
(e) Counting the number of passengers on a bus.
(f) Moving a piece in a game of chess.

3. (a) Give two examples of scalar quantities.
(b) Give two situations involving vector quantities.

4. Using suitable scales, represent these vector quantities by arrows:

(a) A displacement of 20 cm to the left.
(b) A velocity of 10 m per second due East.

(c) A pull of 70 Newtons downwards.
(d) A current of 6 km per hour to the North-West.
(e) A journey of 12 km on a bearing of 130°.

5. State the size and direction of each of the vectors in Fig. 3.

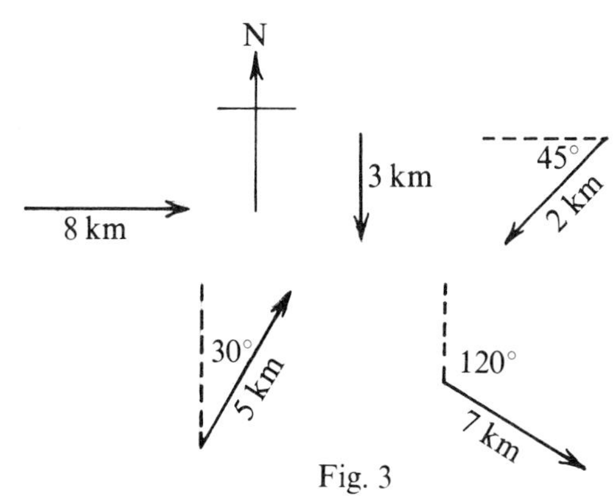

Fig. 3

6. Draw arrows on a convenient scale to represent these vectors:

(a) 14 km in the direction 030°,
(b) 20 km in the direction 290°,
(c) 8 km in the direction 180°.

7. In Fig. 4 arrows A and B represent vectors which have the same size and are in the same direction. Such vectors are said to be *equivalent*. Which other arrows represent equivalent vectors?

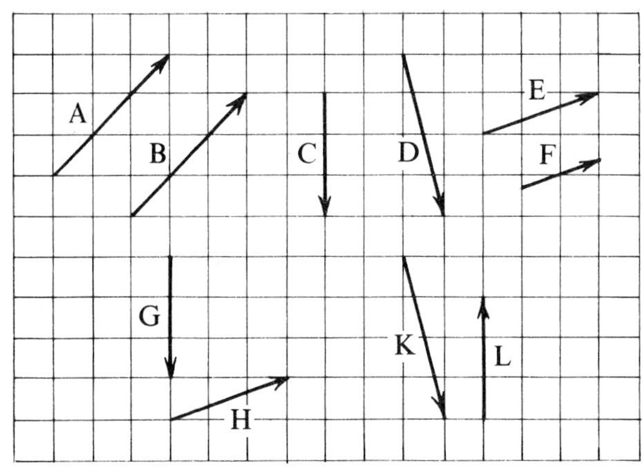

Fig. 4

8. On squared paper draw:
(i) two arrows which represent equivalent vectors,
(ii) two arrows which represent vectors which are not equivalent.

9. On squared paper draw arrows to represent a possible move in a game of chess by:
(i) a king (ii) a pawn (iii) a knight (iv) a bishop (v) a queen.

10. Explain the difference between a scalar quantity and a vector quantity.

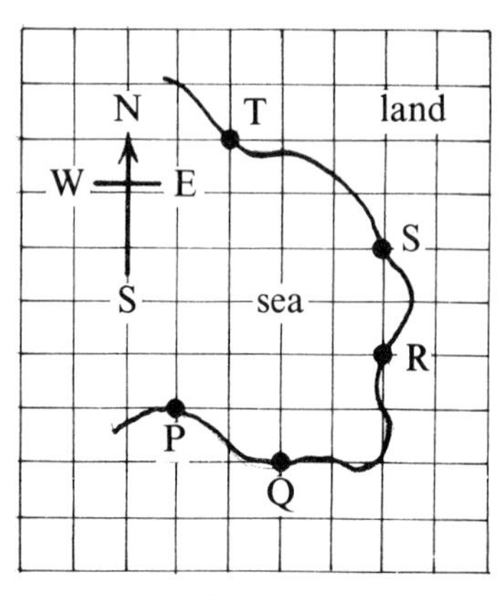

Fig. 5

Fig. 5 represents a coastline with small fishing ports at P, Q, R, S and T. The square grid is in kilometre units.

Port R is 4 km East of P and 1 km North of P. The journey by boat from P to R can be described as 4 km East and 1 km North. This does not mean that the boat first sails 4 km East and then 1 km North, but that at the end of the journey the point reached is 4 km E and 1 km N of P. We can write the movement of the boat as $\begin{pmatrix}4\\1\end{pmatrix}$. This is called a *column vector*. Notice that the top number is the distance to the East and the bottom number is the distance to the North. $\begin{pmatrix}1\\4\end{pmatrix}$ is a different movement. A short way of writing 'the movement from P to R' is $\overrightarrow{PR}$ and so we can write $\overrightarrow{PR} = \begin{pmatrix}4\\1\end{pmatrix}$.

From P to Q the boat travels 2 km East and 1 km South and so $\overrightarrow{PQ} = \begin{pmatrix} 2 \\ -1 \end{pmatrix}$. Notice that -1 shows that the distance of 1 km is in the opposite direction to North.

Check that $\overrightarrow{RP} = \begin{pmatrix} -4 \\ -1 \end{pmatrix}$ and $\overrightarrow{QP} = \begin{pmatrix} -2 \\ 1 \end{pmatrix}$.

Exercise 79

1. (i) Using Fig. 5, give these journeys as column vectors:

$\overrightarrow{PT}$, $\overrightarrow{QR}$, $\overrightarrow{TR}$, $\overrightarrow{QT}$, $\overrightarrow{RS}$

(ii) Using Fig. 5, give the journeys represented by each of the following column vectors:

$\begin{pmatrix} 4 \\ 1 \end{pmatrix}$, $\begin{pmatrix} 2 \\ 2 \end{pmatrix}$, $\begin{pmatrix} -3 \\ 2 \end{pmatrix}$, $\begin{pmatrix} 1 \\ 5 \end{pmatrix}$, $\begin{pmatrix} 1 \\ -6 \end{pmatrix}$.

2. Fig. 6 represents another coastline:

(i) Copy it on squared paper and draw arrows representing the journeys $\overrightarrow{XZ}$, $\overrightarrow{XY}$, $\overrightarrow{RT}$, $\overrightarrow{RZ}$ and $\overrightarrow{ZT}$.

(ii) Write these journeys as column vectors.

(iii) Name the journeys:

$\begin{pmatrix} -1 \\ 3 \end{pmatrix}$, $\begin{pmatrix} -4 \\ -3 \end{pmatrix}$, $\begin{pmatrix} -4 \\ 7 \end{pmatrix}$ and $\begin{pmatrix} -1 \\ 4 \end{pmatrix}$

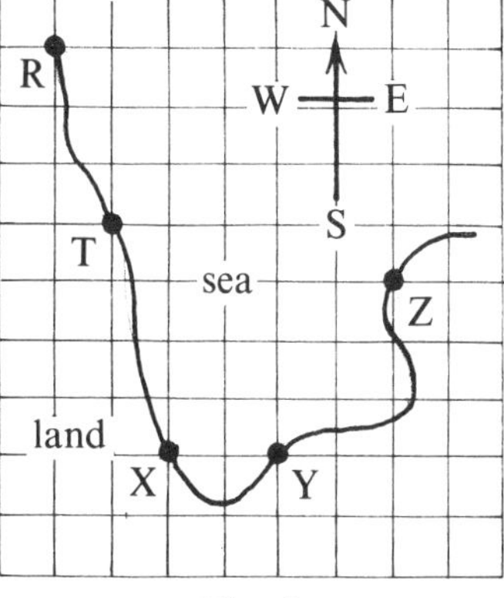

Fig. 6

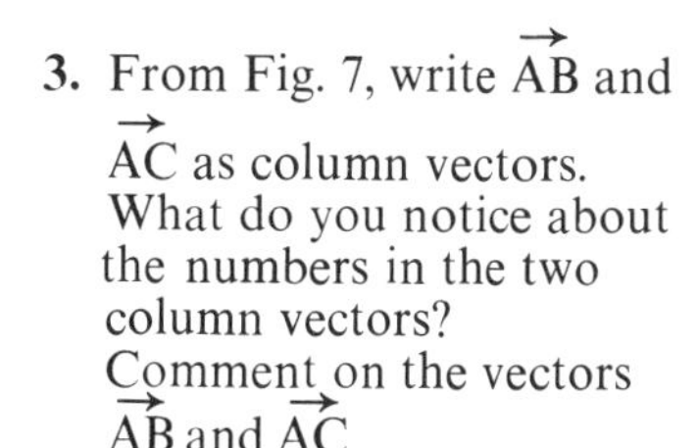

3. From Fig. 7, write $\overrightarrow{AB}$ and $\overrightarrow{AC}$ as column vectors. What do you notice about the numbers in the two column vectors? Comment on the vectors $\overrightarrow{AB}$ and $\overrightarrow{AC}$

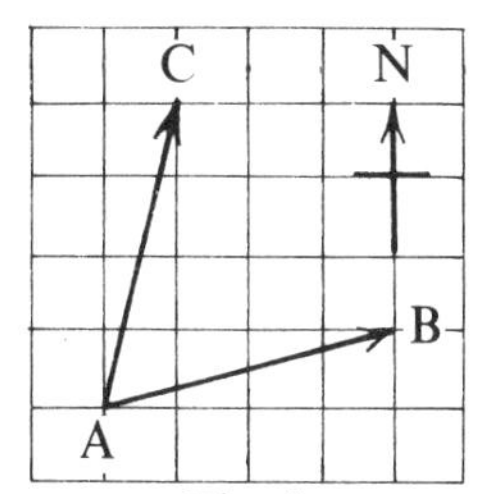

Fig. 7

4. In Fig. 8, P is the point (4, 2).

The vector $\overrightarrow{OP}$ can be regarded as having two parts: an x part of 4 units and a y part of 2 units.

We write $\overrightarrow{OP} = \begin{pmatrix} 4 \\ 2 \end{pmatrix}$. Copy Fig. 8.

Draw arrows to represent the vectors $\overrightarrow{OQ}$, $\overrightarrow{OR}$, $\overrightarrow{OS}$ and $\overrightarrow{OT}$ and write them as column vectors.

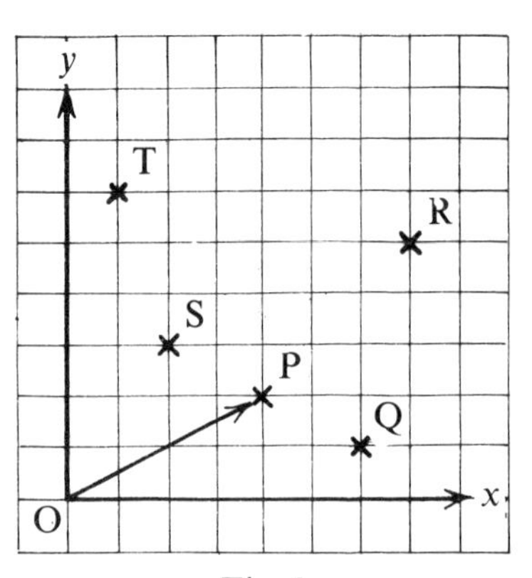

Fig. 8

5. In Fig. 9, $\overrightarrow{OA} = \begin{pmatrix} 4 \\ 1 \end{pmatrix}$ and also $\overrightarrow{BC} = \begin{pmatrix} 4 \\ 1 \end{pmatrix}$.

They are equivalent vectors. Another way of describing these vectors is to say that each represents a translation of 4 units to the right and 1 unit up the page. A is the image of O and C is the image of B.

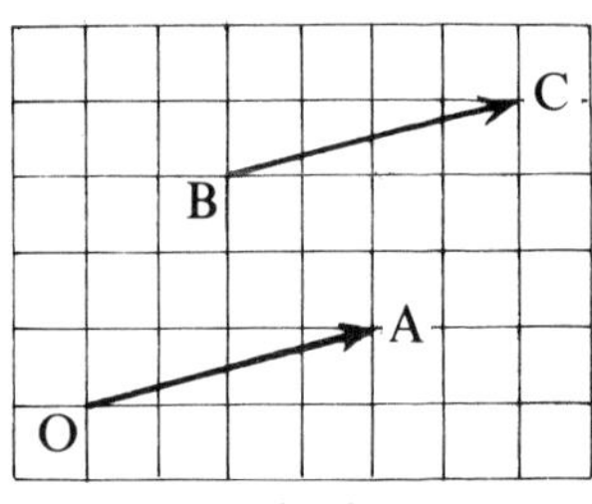

Fig. 9

(i) Copy the diagram.
(ii) Is OA parallel to BC? Is OA equal in length to BC?
(iii) Draw vectors OB and AC. Write each as a column vector. What do you notice? Make two statements about OB and AC.
(iv) What shape is the quadrilateral OABC?

6. (i) Using Fig. 10, write as column vectors $\overrightarrow{OP}$, $\overrightarrow{OQ}$, $\overrightarrow{OR}$, $\overrightarrow{ST}$ and $\overrightarrow{SR}$.

(ii) Name two vectors equal to $\begin{pmatrix} 3 \\ -1 \end{pmatrix}$.

(iii) Write as column vectors $\overrightarrow{RQ}$, $\overrightarrow{QT}$, $\overrightarrow{TS}$ and $\overrightarrow{SP}$.

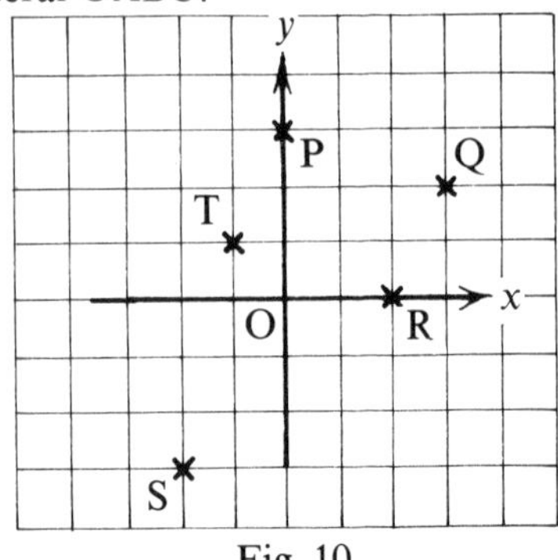

Fig. 10

7. In Fig. 10, if T $(-1, 1)$ is given a translation of $\begin{pmatrix} 2 \\ -3 \end{pmatrix}$, its image, T′, is $(1, -2)$. Find the images of P, Q, R and S.

COMBINING VECTORS

In Fig. 11, $\overrightarrow{GH} = \begin{pmatrix} 3 \\ 1 \end{pmatrix}$ and $\overrightarrow{HK} = \begin{pmatrix} 1 \\ 2 \end{pmatrix}$.

Vector $\overrightarrow{GH}$ moves an object from G to H.

Vector $\overrightarrow{HK}$ moves it from H to K.

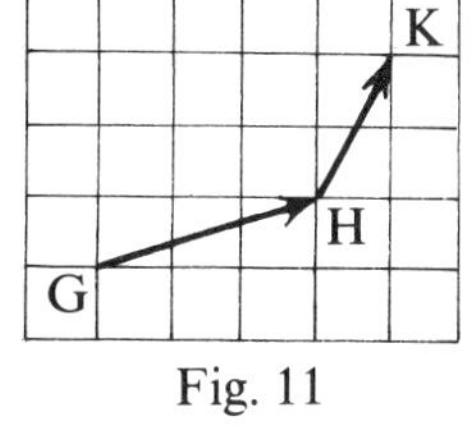

Fig. 11

The result of the two vectors is to move the object from G to K. The vector $\overrightarrow{GK}$ has this same result.

We write $\overrightarrow{GH} + \overrightarrow{HK} = \overrightarrow{GK}$

That is $\begin{pmatrix} 3 \\ 1 \end{pmatrix} + \begin{pmatrix} 1 \\ 2 \end{pmatrix} = \begin{pmatrix} 3+1 \\ 1+2 \end{pmatrix} = \begin{pmatrix} 4 \\ 3 \end{pmatrix}$

We say that $\overrightarrow{GK}$ is the sum of $\overrightarrow{GH}$ and $\overrightarrow{HK}$.

Exercise 80

1. Draw diagrams on squared paper to show that:

(i) $\begin{pmatrix} 2 \\ 1 \end{pmatrix} + \begin{pmatrix} 3 \\ 1 \end{pmatrix} = \begin{pmatrix} 5 \\ 2 \end{pmatrix}$ (ii) $\begin{pmatrix} 1 \\ 4 \end{pmatrix} + \begin{pmatrix} 2 \\ 3 \end{pmatrix} = \begin{pmatrix} 3 \\ 7 \end{pmatrix}$

(iii) $\begin{pmatrix} 2 \\ 2 \end{pmatrix} + \begin{pmatrix} 4 \\ 2 \end{pmatrix} = \begin{pmatrix} 6 \\ 4 \end{pmatrix}$.

2. Find these vector sums:

(i) $\begin{pmatrix} 5 \\ 2 \end{pmatrix} + \begin{pmatrix} 1 \\ 4 \end{pmatrix}$ (ii) $\begin{pmatrix} 3 \\ 7 \end{pmatrix} + \begin{pmatrix} -2 \\ 1 \end{pmatrix}$

(iii) $\begin{pmatrix} 4 \\ 4 \end{pmatrix} + \begin{pmatrix} 0 \\ -1 \end{pmatrix}$ (iv) $\begin{pmatrix} -2 \\ -3 \end{pmatrix} + \begin{pmatrix} 6 \\ 9 \end{pmatrix}$.

3. Draw diagrams on squared paper to show that:

(i) $\begin{pmatrix}3\\2\end{pmatrix}+\begin{pmatrix}-1\\0\end{pmatrix}=\begin{pmatrix}2\\2\end{pmatrix}$ (ii) $\begin{pmatrix}4\\1\end{pmatrix}+\begin{pmatrix}-2\\-1\end{pmatrix}=\begin{pmatrix}2\\0\end{pmatrix}$

(iii) $\begin{pmatrix}5\\-2\end{pmatrix}+\begin{pmatrix}3\\3\end{pmatrix}=\begin{pmatrix}8\\1\end{pmatrix}$.

4. A column vector is often denoted by a single small letter, underlined, such as $\underline{a}$. Let

$$\underline{a}=\begin{pmatrix}2\\4\end{pmatrix},\quad \underline{b}=\begin{pmatrix}3\\5\end{pmatrix},\quad \underline{c}=\begin{pmatrix}-2\\1\end{pmatrix}\quad\text{and}\quad \underline{d}=\begin{pmatrix}4\\-5\end{pmatrix}$$

State as column vectors:

(i) $\underline{a}+\underline{b}$ (ii) $\underline{b}+\underline{c}$ (iii) $\underline{c}+\underline{a}$
(iv) $\underline{b}+\underline{a}$ (v) $\underline{d}+\underline{b}$ (vi) $\underline{d}+\underline{c}$

5. Solve the following equations:

(i) $\begin{pmatrix}4\\1\end{pmatrix}+\begin{pmatrix}-3\\2\end{pmatrix}=\begin{pmatrix}x\\y\end{pmatrix}$ (ii) $\begin{pmatrix}2\\5\end{pmatrix}+\begin{pmatrix}6\\-3\end{pmatrix}=\begin{pmatrix}x\\y\end{pmatrix}$

(iii) $\begin{pmatrix}7\\-2\end{pmatrix}+\begin{pmatrix}-4\\-1\end{pmatrix}=\begin{pmatrix}x\\y\end{pmatrix}$ (iv) $\begin{pmatrix}x\\y\end{pmatrix}+\begin{pmatrix}5\\8\end{pmatrix}=\begin{pmatrix}6\\7\end{pmatrix}$

(v) $\begin{pmatrix}-2\\-1\end{pmatrix}+\begin{pmatrix}x\\y\end{pmatrix}=\begin{pmatrix}3\\0\end{pmatrix}$ (vi) $\begin{pmatrix}x\\y\end{pmatrix}+\begin{pmatrix}1\\-3\end{pmatrix}=\begin{pmatrix}5\\7\end{pmatrix}$.

6. In Fig. 12, $\underline{r}=\begin{pmatrix}4\\2\end{pmatrix}$ and $\underline{t}=\begin{pmatrix}8\\4\end{pmatrix}$

$$\underline{r}+\underline{r}=\begin{pmatrix}4\\2\end{pmatrix}+\begin{pmatrix}4\\2\end{pmatrix}$$

$$=\begin{pmatrix}4+4\\2+2\end{pmatrix}=\begin{pmatrix}2\times 4\\2\times 2\end{pmatrix}$$

$$=\begin{pmatrix}8\\4\end{pmatrix}=\underline{t}.$$

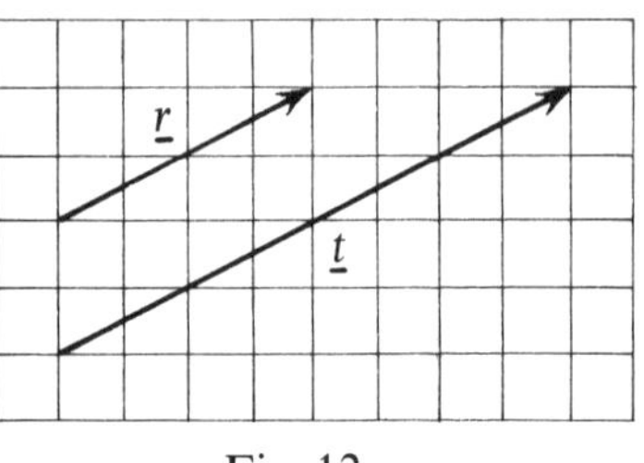

Fig. 12

We write $2\underline{r}$ for $\underline{r}+\underline{r}$ and so $2\underline{r}=\underline{t}$.

$$3\underline{r} = \underline{r} + \underline{r} + \underline{r} = \begin{pmatrix}4\\2\end{pmatrix} + \begin{pmatrix}4\\2\end{pmatrix} + \begin{pmatrix}4\\2\end{pmatrix} = \begin{pmatrix}4+4+4\\2+2+2\end{pmatrix}$$
$$= \begin{pmatrix}3\times 4\\3\times 2\end{pmatrix} = \begin{pmatrix}12\\6\end{pmatrix}$$

Similarly, $5\underline{r} = \begin{pmatrix}5\times 4\\5\times 2\end{pmatrix} = \begin{pmatrix}20\\10\end{pmatrix}$.

State as column vectors: $4\underline{r}$, $7\underline{r}$ and $\frac{1}{2}\underline{r}$.

7. Let $\underline{u} = \begin{pmatrix}3\\4\end{pmatrix}$, $\underline{w} = \begin{pmatrix}6\\3\end{pmatrix}$ and $\underline{x} = \begin{pmatrix}12\\-8\end{pmatrix}$. State as column vectors:

$2\underline{u}$, $3\underline{w}$, $4\underline{x}$, $\frac{1}{3}\underline{w}$, $\frac{1}{2}\underline{x}$, $2\underline{u} + \underline{x}$ and $3\underline{w} + 3\underline{u}$.

8. In Fig. 13, $\underline{p} = \begin{pmatrix}5\\3\end{pmatrix}$
and $\underline{s} = \begin{pmatrix}-5\\-3\end{pmatrix}$

$\underline{p} + \underline{s} = \begin{pmatrix}5\\3\end{pmatrix} + \begin{pmatrix}-5\\-3\end{pmatrix}$

$= \begin{pmatrix}0\\0\end{pmatrix} = \underline{0}$

$\underline{0}$ or $\begin{pmatrix}0\\0\end{pmatrix}$ is called the *zero vector*.

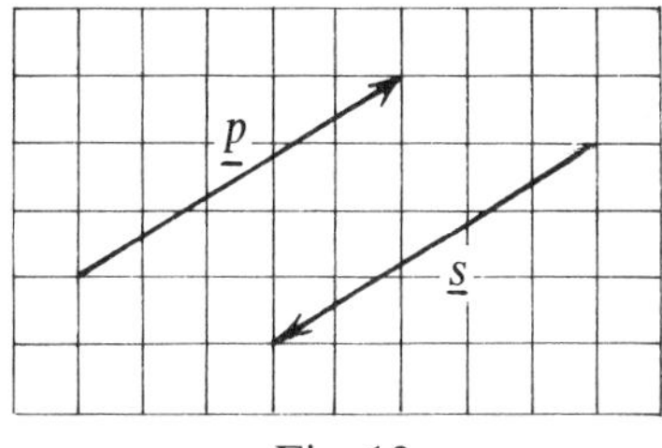

Fig. 13

We say that $\underline{p}$ is the *inverse* of $\underline{s}$ for addition and that $\underline{s}$ is the inverse of $\underline{p}$ for addition.

The inverse of $\underline{p}$ can be written $-\underline{p}$.

If $\underline{h} = \begin{pmatrix}3\\-8\end{pmatrix}$, then $-\underline{h} = \begin{pmatrix}-3\\8\end{pmatrix}$.

(i) Copy and complete $\begin{pmatrix}2\\5\end{pmatrix} + \begin{pmatrix} \\ \end{pmatrix} = \begin{pmatrix}0\\0\end{pmatrix}$, $\begin{pmatrix}-3\\1\end{pmatrix} + \begin{pmatrix} \\ \end{pmatrix} = \begin{pmatrix}0\\0\end{pmatrix}$,

and $\begin{pmatrix}4\\-6\end{pmatrix} + \begin{pmatrix} \\ \end{pmatrix} = \begin{pmatrix}0\\0\end{pmatrix}$.

(ii) State the inverses of $\begin{pmatrix}2\\5\end{pmatrix}$, $\begin{pmatrix}-3\\1\end{pmatrix}$ and $\begin{pmatrix}4\\-6\end{pmatrix}$.

9. Using the vectors of Question **7**, state as column vectors
$-\underline{u}$, $-\underline{w}$ and $-\underline{x}$

10. (i) Draw each of the following vectors:

$$\begin{pmatrix}5\\1\end{pmatrix}, \begin{pmatrix}3\\2\end{pmatrix}, \begin{pmatrix}-1\\4\end{pmatrix}, \begin{pmatrix}6\\6\end{pmatrix}, \begin{pmatrix}0\\-2\end{pmatrix}, \begin{pmatrix}4\\0\end{pmatrix}.$$

(ii) State and draw their inverses.

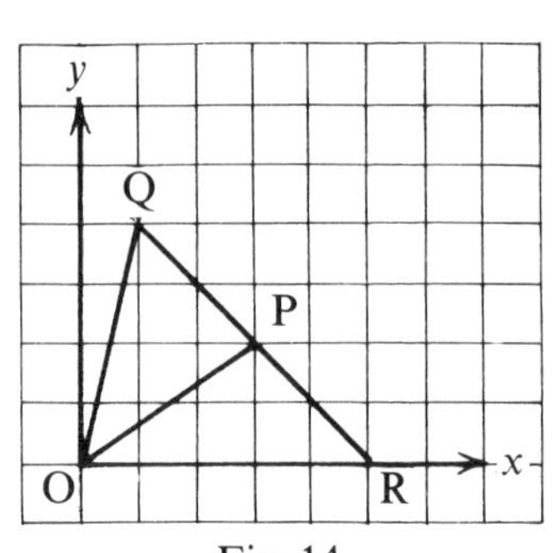

Fig. 14

In Fig. 14, $\vec{OP} = \begin{pmatrix}3\\2\end{pmatrix}$

$5\vec{OP} = 5\begin{pmatrix}3\\2\end{pmatrix} = \begin{pmatrix}5 \times 3\\5 \times 2\end{pmatrix} = \begin{pmatrix}15\\10\end{pmatrix}.$

$\vec{PQ} = \begin{pmatrix}-2\\2\end{pmatrix}$ and $\vec{PR} = \begin{pmatrix}2\\-2\end{pmatrix}.$

Note that $\vec{PQ}$ and $\vec{PR}$ are inverse vectors since they have the same length but opposite directions.

$\vec{OP} + \vec{PQ} = \begin{pmatrix}3\\2\end{pmatrix} + \begin{pmatrix}-2\\2\end{pmatrix} = \begin{pmatrix}1\\4\end{pmatrix} = \vec{OQ}$ and

$\vec{OP} + \vec{PR} = \begin{pmatrix}3\\2\end{pmatrix} + \begin{pmatrix}2\\-2\end{pmatrix} = \begin{pmatrix}5\\0\end{pmatrix} = \vec{OR}$

Exercise 81

1. Using Fig. 14, write each of the following as a single column vector: $\vec{OQ}$, $\vec{OR}$, $\vec{OQ} + \vec{OR}$, $\vec{OP}$ and $2\,\vec{OP}$. You should find that $\vec{OQ} + \vec{OR} = 2\,\vec{OP}$.
2. On squared paper label axes Ox and Oy and two convenient points Q and R. Mark P as the mid-point of QR. Using column vectors show that $\vec{OQ} + \vec{OR} = 2\vec{OP}$.
3. Look back at Fig. 11. We found that $\vec{GH} + \vec{HK} = \vec{GK}$.

   In Fig. 15, $\vec{OB} = \vec{OA} + \vec{AB}$

   and $\vec{OC} = \vec{OB} + \vec{BC}$.

   Hence $\vec{OC} = \vec{OB} + \vec{BC}$

   $= (\vec{OA} + \vec{AB}) + \vec{BC}$.

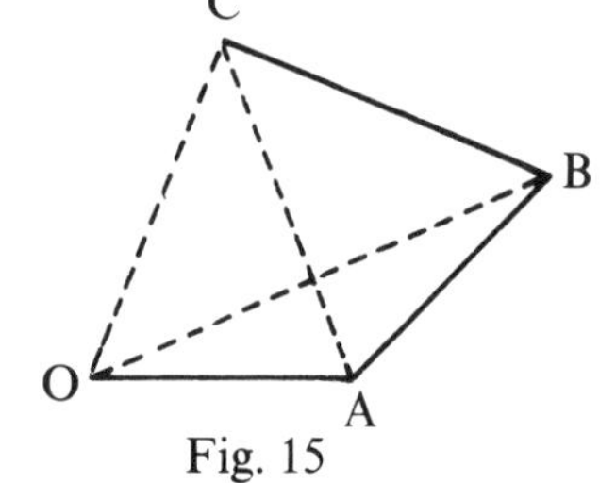

Fig. 15

Copy and complete: $\overrightarrow{OC} = \ldots + \overrightarrow{AC} = \ldots + (\ldots + \ldots)$

and so $(\overrightarrow{OA} + \overrightarrow{AB}) + \ldots = \ldots + (\ldots + \ldots)$

4. In Fig. 16, $\underline{a} = \begin{pmatrix} 4 \\ 0 \end{pmatrix}$, $\underline{b} = \begin{pmatrix} 3 \\ 3 \end{pmatrix}$ and $\underline{c} = \begin{pmatrix} -5 \\ 2 \end{pmatrix}$.

Write as single column vectors:

(i) $\underline{a} + \underline{b}$ (ii) $(\underline{a} + \underline{b}) + \underline{c}$
(iii) $\underline{b} + \underline{c}$ (iv) $\underline{a} + (\underline{b} + \underline{c})$.

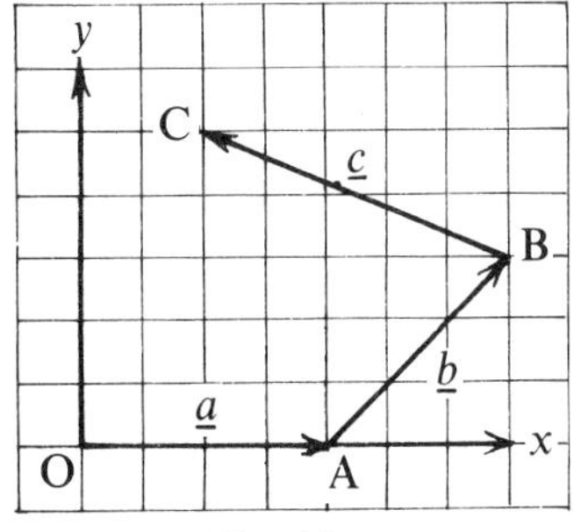

Fig. 16

You should find that $(\underline{a} + \underline{b}) + \underline{c} = \underline{a} + (\underline{b} + \underline{c})$.

5. Copy Fig. 16. Draw an arrow from O to B and label it $\underline{a} + \underline{b}$. Draw an arrow from O to C and label is $(\underline{a} + \underline{b}) + \underline{c}$. Draw a separate diagram to show $\underline{b} + \underline{c}$ and $\underline{a} + (\underline{b} + \underline{c})$.

6. Choose your own vectors $\underline{p}$, $\underline{q}$ and $\underline{r}$. Draw a diagram to show $(\underline{p} + \underline{q}) + \underline{r}$ and another to show $\underline{p} + (\underline{q} + \underline{r})$.

7. In Fig. 17, P′ is (6, 0), Q′ is (4, 4). P is the mid-point of OP′ and Q is the mid-point of OQ′. State the coordinates of P and Q.

Give $\overrightarrow{PQ}$ and $\overrightarrow{P'Q'}$ as column vectors.

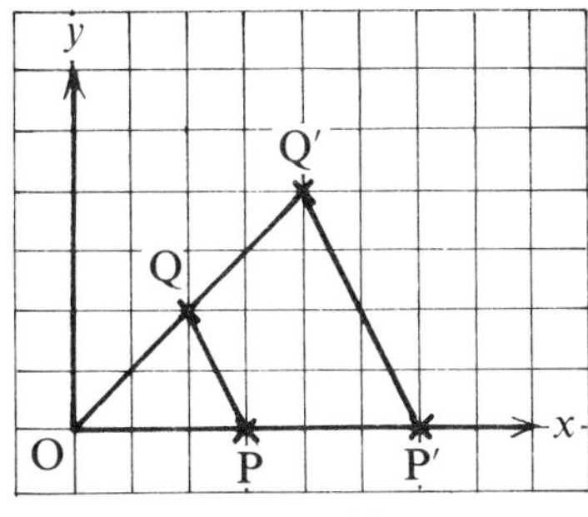

Fig. 17

What is the relation between these vectors?

What can you say about the lengths of $\overrightarrow{PQ}$ and $\overrightarrow{P'Q'}$? What can you say about the directions of $\overrightarrow{PQ}$ and $\overrightarrow{P'Q'}$?

8. Use the idea of enlargement to give another explanation as to why the length of PQ in Fig. 17 is half the length of P′Q′ and why PQ is parallel to P′Q′.

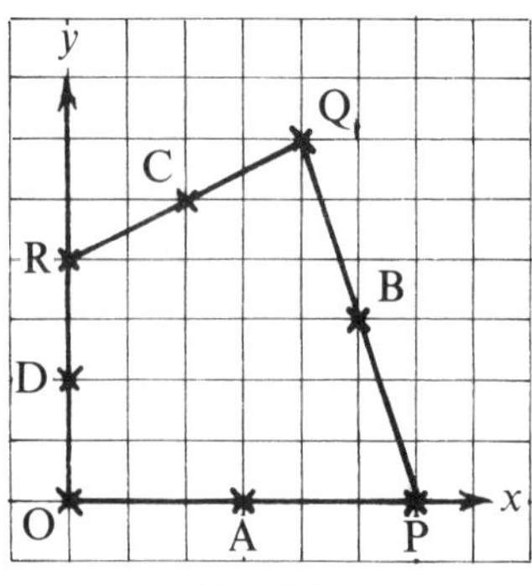

Fig. 18

9. In Fig. 18, P is (6, 0), Q is (4, 6) and R is (0,4). A is the mid-point of OP, B is the mid-point of PQ, C is the mid-point of QR and D is the mid-point of OR. State the coordinates of A, B, C and D. Give $\overrightarrow{AB}$ and $\overrightarrow{DC}$ as column vectors.

What can you say about their lengths and directions? What is the shape of quadrilateral ABCD? Show that $\overrightarrow{AB} = \frac{1}{2}\overrightarrow{OQ}$.

10. Using Fig. 18, show by column vectors that $\overrightarrow{BC} = \overrightarrow{AD} = \frac{1}{2}\overrightarrow{PR}$.

11. On squared paper, draw diagrams to show that

(i) $\begin{pmatrix}5\\0\end{pmatrix} + \begin{pmatrix}-2\\4\end{pmatrix} + \begin{pmatrix}-3\\-4\end{pmatrix} = \underline{0}$

(ii) $\begin{pmatrix}2\\3\end{pmatrix} + \begin{pmatrix}4\\-1\end{pmatrix} + \begin{pmatrix}-8\\-1\end{pmatrix} + \begin{pmatrix}2\\-1\end{pmatrix} = \underline{0}$

Why do you return to the starting point in each case?

12. Solve these vector equations:

(i) $\begin{pmatrix}x\\y\end{pmatrix} + \begin{pmatrix}4\\-7\end{pmatrix} = \begin{pmatrix}1\\0\end{pmatrix}$ (ii) $\begin{pmatrix}2\\5\end{pmatrix} + \begin{pmatrix}x\\y\end{pmatrix} = \begin{pmatrix}3\\4\end{pmatrix}$

(iii) $\begin{pmatrix}x\\y\end{pmatrix} = 2\begin{pmatrix}4\\1\end{pmatrix} + 3\begin{pmatrix}1\\2\end{pmatrix}$ (iv) $3\begin{pmatrix}x\\y\end{pmatrix} + \begin{pmatrix}-5\\1\end{pmatrix} = \begin{pmatrix}4\\7\end{pmatrix}$.

13. Draw diagrams to show that:

(i) $\begin{pmatrix}2\\0\end{pmatrix} + \begin{pmatrix}4\\1\end{pmatrix} + \begin{pmatrix}1\\3\end{pmatrix} = \begin{pmatrix}7\\4\end{pmatrix}$

(ii) $\begin{pmatrix}-1\\-1\end{pmatrix} + \begin{pmatrix}-2\\5\end{pmatrix} + \begin{pmatrix}6\\-2\end{pmatrix} = \begin{pmatrix}3\\2\end{pmatrix}$.

20 · QUADRATIC EQUATIONS BY FACTORISING

Exercise 82

1. Copy and complete the following table:

a	4	1	0	8	7	-3	0	$1\frac{1}{2}$	2.3	0
b	3	6	5	0	1	2	-4	6	0	0
ab	12			0						

2. Copy and complete:

$7 \times \ldots = 14$, $\ldots \times 5 = 20$, $\ldots \times 9 = 9$, $3 \times \ldots = 0$,
$5 \times \ldots = 0$, $\ldots \times 7 = 0$, $3 \times \ldots = 1$, $\ldots \times \frac{3}{4} = 1$,
$5 \times \ldots = 18$, $\ldots \times \frac{2}{3} = 0$, $5.6 \times \ldots = 0$, $0.5 \times \ldots = 2$.

3. Make a statement about n for each of the following:

(i) $5n = 15$ (ii) $7n = 0$ (iii) $9n = 1$
(iv) $0 \times n = 0$ (v) $n \times 7 = 21$ (vi) $n \times 0 = 0$
(vii) $n \times 4 = 3$ (viii) $n \times 16 = 0$ (ix) $n \times 0 = n$

4. You are told that $ab = 0$. What follows:

(i) if $b = 7$ (ii) if $a = 5$ (iii) if $a = 0$?

5. State the value of $(x - 2)(x - 5)$ if $x =$

(i) 8 (ii) 5 (iii) 3 (iv) 2 (v) 0

6. Find the value of $(x - 3)(x - 6)$ for each of the whole number values of x from 1 to 7.

If $(x - 3)(x - 6) = 0$, what are the possible values of x?

7. Find the value of $(x - 2)(x - 4)$ for $x = 4.5$ and for $x = 3.2$.

8. State the value of $x(x - 3)$ for $x = 0, 1, 2$ and 3.

9. State the value of $(x + 2)(x + 4)$ for whole number values of x from -5 to 0.

If $(x + 2)(x + 4) = 0$, what are the possible values of x?

10. State the value of $(x - 3)(x + 4)$ for $x = 4, 3, 2, -2, -3$ and -4.
If $(x - 3)(x + 4) = 0$, what are the possible values of x?

State the possible values of x for each of the following equations:

11. $(x - 7)(x - 5) = 0$

12. $(x + 6)(x + 2) = 0$

13. $(x + 3)(x - 4) = 0$

14. $x(x - 8) = 0$

15. When $x = -4$,

$$x^2 - x - 6 = (-4)^2 - (-4) - 6 = 16 + 4 - 6 = 14$$

In this way, find the value of $x^2 - x - 6$ for $x = -3, -2, -1, 0,$ $1, 2, 3$ and 4. If $x^2 - x - 6 = 0$, what values of x are possible?

16. By trying $x = -2, -1, 0, 1$ and 2, find the values of x for which $x^2 - x - 2 = 0$.

The above exercise shows that

if $ab = 0$, then either $a = 0$ or $b = 0$
(or both a and b are 0)

Hence if $(x - 2)(x + 7) = 0$

either $x - 2 = 0$ or $x + 7 = 0$

and so $x = 2$ or $x = -7$

EXAMPLE 1: *Solve* $x^2 - 8x + 15 = 0$

Factorising the left-hand side of the equation,

$$(x - 5)(x - 3) = 0$$

Either $x - 5 = 0$ or $x - 3 = 0$

$\therefore$ $x = 5$ or $x = 3$

EXAMPLE 2: *Solve* $y^2 - 7y = 18$

Subtracting 18 from each side, we have

$$y^2 - 7y - 18 = 0$$
$$(y - 9)(y + 2) = 0$$

Either $y - 9 = 0$ or $y + 2 = 0$

$\therefore$ $y = 9$ or $y = -2$

Exercise 83

Solve the following equations:

1. $x^2 - 7x + 10 = 0$ **2.** $x^2 - 4x + 3 = 0$ **3.** $x^2 + 7x + 12 = 0$
4. $x^2 + 7x + 6 = 0$ **5.** $x^2 - 3x - 10 = 0$ **6.** $x^2 + 2x - 3 = 0$
7. $x^2 - 9x = 0$ **8.** $x^2 - 9 = 0$ **9.** $x^2 + 2x - 35 = 0$
10. $x^2 - 2x - 15 = 0$ **11.** $y^2 - y - 20 = 0$ **12.** $y^2 + y - 30 = 0$
13. $10 - 3u - u^2 = 0$ **14.** $12 + 4n - n^2 = 0$

15. $x^2 - 7x = 8$ (First subtract 8 from each side)

16. $y^2 + 5y = 14$

17. $3x^2 - 9x + 6 = 0$ (First divide all terms by 3)

18. $5y^2 + 5y = 30$ (First divide each side by 5).

19. Find the value of $x^2 - 2x - 8$ for each of the following values of x: 5, 4, 3, 2, 1, 0, -1, -2, -3. Use your results to state the solutions of $x^2 - 2x - 8 = 0$.
Also solve the equation by factorising.

20. Find the value of $x^2 - 10x + 24$ for each of the whole number values of x from 2 to 8. Hence state the solutions of $x^2 - 10x + 24 = 0$.
Also solve the equation by factorising.

21. Has $x^2 - 3x + 1$ any factors? Are there any whole number values of x which satisfy $x^2 - 3x + 1 = 0$?

22. Expand $(x + 3)^2$ and hence solve the equation $(x + 3)^2 - 4 = 0$.

Solve the following equations:

23. $(x - 5)^2 - 2x + 2 = 0$ **24.** $(2y + 3)(y + 1) = (y + 3)^2$
25. $(y - 1)^2 = (y - 5)^2$ **26.** $(n - 2)(n + 3) = 14$
27. $(p + 3)(p + 2) = 2$.

PROBLEMS USING QUADRATIC EQUATIONS

EXAMPLE: *I write down two numbers. One of them is 2 larger than the other. I square each number and add the answers together. I obtain 202. What were the original numbers*

Let the smaller number be n.

Then the other number is $n + 2$.

The sum of the squares is $n^2 + (n + 2)^2$ and this is 202.

$$\begin{aligned} \text{Hence} \quad n^2 + (n+2)^2 &= 202 \\ n^2 + n^2 + 4n + 4 &= 202 \\ 2n^2 + 4n - 198 &= 0 \\ n^2 + 2n - 99 &= 0 \\ (n-9)(n+11) &= 0 \\ n &= 9 \text{ or } -11. \end{aligned}$$

Taking $n = 9$, the numbers were 9 and 11.

Check: $9^2 + 11^2 = 81 + 121 = 202$

Taking $n = -11$, the numbers were -11 and -9.

Exercise 84

1. I write down two numbers. One is 3 larger than the other. I square each number and add the answers together. I obtain 65. Let the smaller number be n. Form an equation for n and simplify it. Solve the equation and so find the two original numbers.

2. I write down two numbers. One is three times the other. When I square each number and add the answers together I get 160. Let the smaller number be n. Form an equation for n and use it to find the two original numbers.

3. The length of a rectangle is 6 metres greater than the width. The area is 160 square metres. Let the width be x metres. Write down an expression for the length and an expression for the area. Put this equal to 160 and solve the equation. State the length and width.

4. John is 4 years older than his sister, Jane. The product of their ages is 96. Let Jane's age be n years. Write down an expression for John's age and an expression for the product of their ages. Form an equation and solve it to find the two ages.

5. Triangle ABC has a right-angle at A. AC is 7 cm longer than AB; BC is 2 cm longer than AC. Let the length of AB be y cm. Write down expressions for the lengths of AC and BC. Form an equation by using Pythagoras' Theorem. Solve it and so find the lengths of the three sides of the triangle.

6. A rectangle has a length of x cm and a width of $(x - 7)$ cm. The diagonals are each $(x + 1)$ cm. Form an equation for x and solve it. State the length of the rectangle.

7. A stone is thrown vertically upwards. The formula for its height, h metres, after t seconds is $h = 30t - 5t^2$. Put $h = 25$ and solve the equation obtained. This gives the two times when the height of the stone is 25 m. Why are there two times?

Find also the times at which the height of the stone is 40 m.

8. The diagram shows two graphs which cut at A and B.

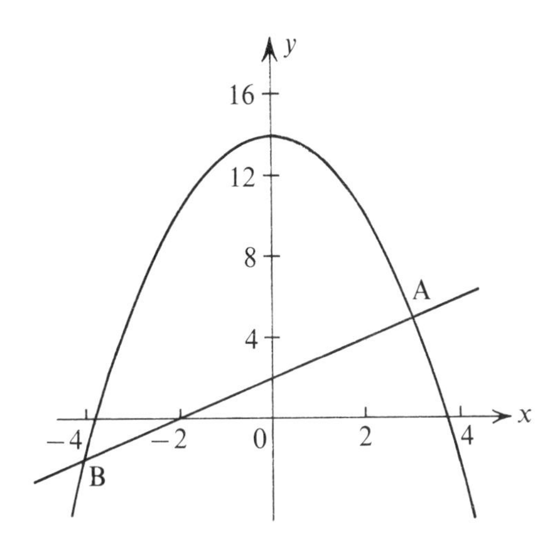

The equation of the straight line is $y = x + 2$ and the equation of the curve is $y = 14 - x^2$.

At A and B the coordinates satisfy both equations.

Show that (1, 3) is on the line but not on the curve.

Show that (2, 10) is on the curve but not on the line.

At A and B, $x + 2 = 14 - x^2$. Solve this equation and so state the coordinates of A and B.

9. Use the method of Question **8** to find the points of intersection of the straight line $y = 4x$ and the curve $y = x^2 - 5$.

10. At the points of intersection of $y = x + 3$ and $xy = 10$, the equation $x(x + 3) = 10$ is satisfied. Solve this equation and so find the two points of intersection.

21 · TRAVEL GRAPHS

DISTANCE-TIME GRAPHS

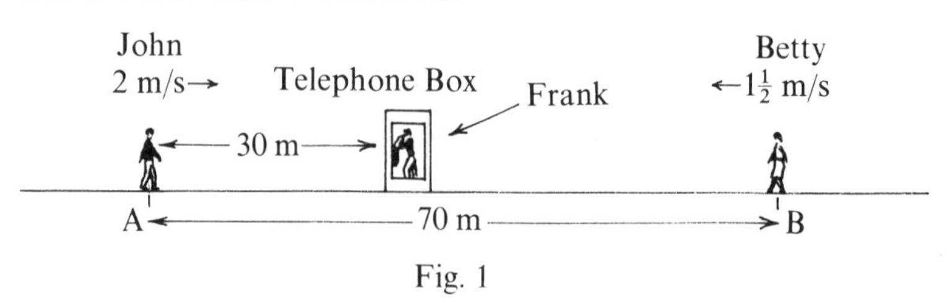

Fig. 1

John passes point A at a brisk walking speed of 2 m/s. At the same moment Betty, who is walking at $1\frac{1}{2}$ m/s in the opposite direction, passes point B. Here is the table for the distances of John and Betty from A at various times.

Time (s)	0	10	20	30	40
John's distance from A (m)	0	20	40	60	80
Betty's distance from A (m)	70	55	40	25	10

Points have been plotted in Fig. 2 for John's distances 0, 20, 40, 60, 80 m at times 0, 10, 20, 30, 40 s. They lie on a straight line because his speed remains the same. This is the travel graph for John. From it we can say how far John is from A at any time up to 40 s. For example, at 25 s he is 50 m from A.

Points have also been plotted for Mary's distances from A. The straight line obtained is Mary's travel graph. Notice that John and Mary are walking in opposite directions. If we take distances and velocities to the right as positive, then John's velocity is $+2$ m/s and Mary's is $-1\frac{1}{2}$ m/s. John's graph goes up to the right and we say that it has a positive gradient. Mary's goes down to the right and we say that it has a negative gradient.

Frank is in the telephone-box (Fig. 1). His velocity is zero. His distance from A remains at 30 m and so his graph is a straight line parallel to the time axis.

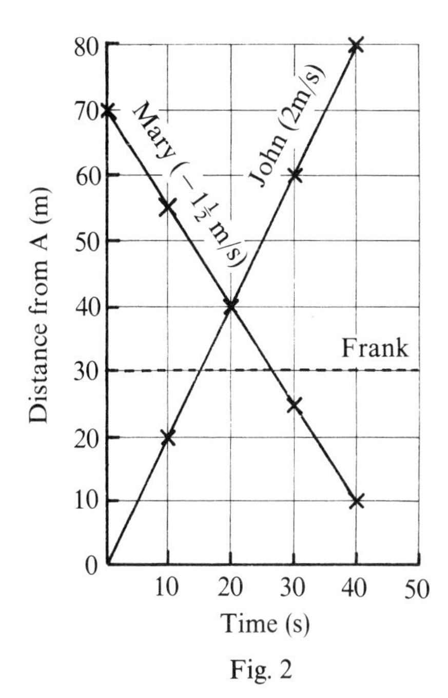

Fig. 2

The graphs for John and Betty cross at the point (20, 40). They pass each other after 20 s at a distance of 40 m from A.

Exercise 85

1. Tom passes point A walking towards B at $1\frac{1}{2}$ m/s and at the same moment Doris passes B walking towards A at 1 m/s. Draw up a table showing their distances from A at 0, 10, 20, 30, 40, 50, 60 seconds. Draw their distance-time graphs using 1 cm for 5 s and 1 cm for 5 m. Add the graph for Frank who is still in the telephone-box.

When do Tom and Doris pass each other and how far are they from A?

When does Doris pass Frank?

2. Draw up a table showing the distances of Tom and Doris from point B at 0, 10, 20, 30, 40, 50, 60 s. Draw the distance-time graphs for the distances from B of Tom, Doris and Frank. Answer the same questions as in Question **1**.

3. Here is a table for Pete who walked from B towards A, waited at the telephone-box and then walked back towards B.

Time (s)	0	10	20	30	40	50	60
Distance from A (m)	70	50	30	30	30	45	60

Draw Pete's distance-time graph.

(i) What was his speed towards A?
(ii) How long did he wait at the telephone-box?
(iii) What was his speed away from A?

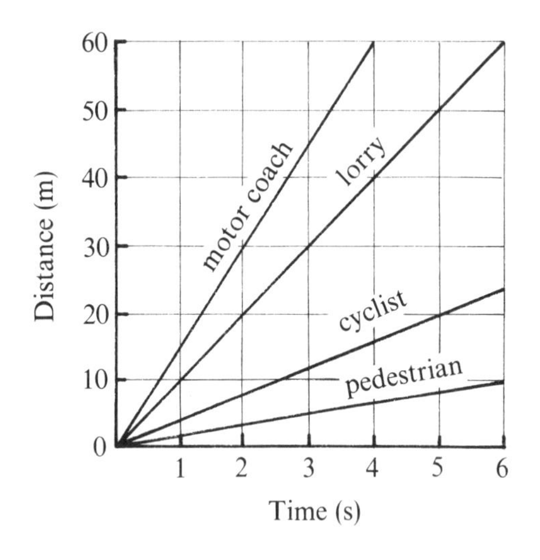

Fig. 3

4. Fig. 3 shows four distance-time graphs:

(i) How far does the coach go in 4 seconds? What is its speed in m/s?
(ii) How far does the cyclist go in 5 seconds? What is his speed in m/s?
(iii) Find the speed of the lorry and of the pedestrian in m/s.
(iv) Calculate the speed of the coach and of the pedestrian in km/h.
(v) The line for a car having a speed of 30 m/s passes through $(1, b)$ and $(2, c)$. What are the values of b and c?

5. Using a time axis for 5 hours and a distance axis for 800 kilometres, draw in one diagram the distance-time graphs for:

(i) an aircraft flying at 600 km/h,
(ii) A train travelling at 150 km/h,
(iii) a skier travelling at 80 km/h,
(iv) a boat travelling at 30 km/h,
(v) a policeman on point duty.

6. Draw graphs to illustrate the following:

(i) A man cycles for 1 h at 15 km/h, stops for 1 h and then continues for 2 h at 10 km/h.
(ii) A boat cruises at 20 km/h for $1\frac{1}{2}$ h and returns to its starting point in 2 h.

7. At 10 a.m. George left Ace Dyke on the River Trout in his motor-boat and cruised 12 km downstream to Beeton where he stopped for a while and then cruised back upstream to Ace Dyke. His distance-time graph is ABCD in Fig. 4.

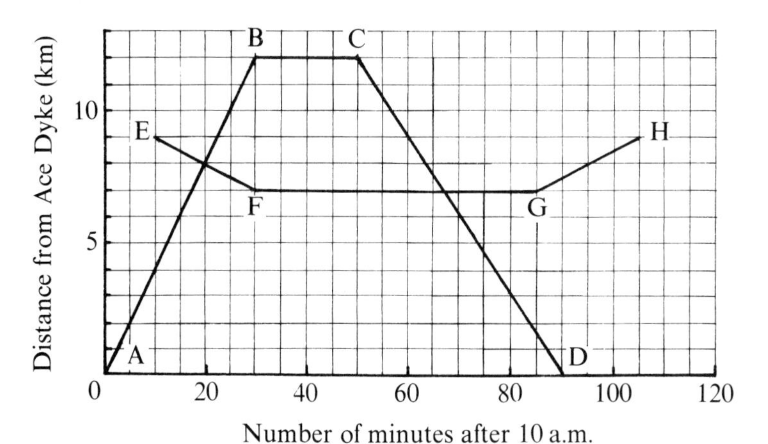

Fig. 4

(i) How long did he take to reach Beeton?
(ii) What was his speed downstream in km/h?
(iii) How long did he spend at Beeton?
(iv) How long did he take for the return journey?
(v) What was his speed upstream?

In Fig. 4 there is also the graph EFGH for a fisherman who left his riverside house in a rowing boat.

(vi) How far was his house from Ace Dyke?
(vii) What time did he leave home?
(viii) At what speed did he row?
(ix) How far from Ace Dyke was he when he fished?
(x) Estimate when George passed the fisherman.

8. In Fig. 5 the continuous line is the graph of a river boat which left A to travel downstream to D, passing through lock B and lock C on the way.

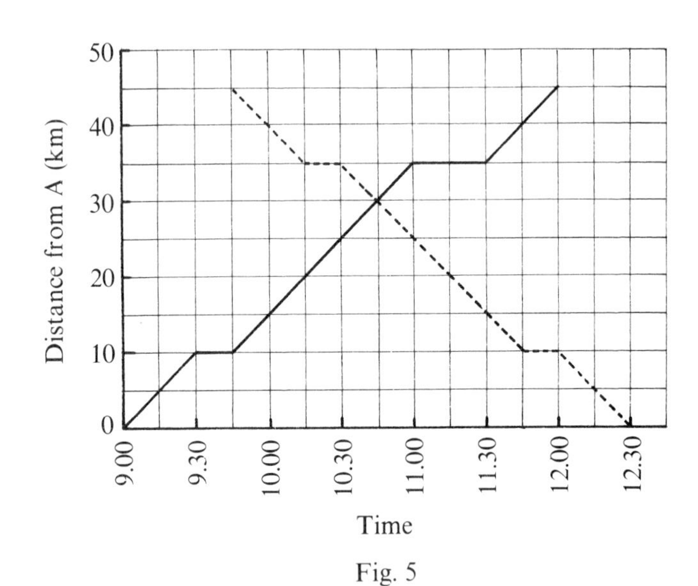

Fig. 5

(i) When did the boat arrive at D?
(ii) How far is D from A?
(iii) How long was the boat at lock B?
(iv) How long was it at lock C?
(v) How far is lock C from D?
(vi) At what speed did the boat travel?

The broken line is the graph of a second boat.

(vii) Did it travel upstream or downstream?
(viii) How long did its journey take?
(ix) What was its speed?
(x) When did it pass the first boat?

9. A cyclist started from home at 11.00 to cycle towards the coast which was 40 km away. He cycled at a steady speed of 16 km/h but stopped for lunch from 12.15 until 12.45. At what time did he arrive at the coast? Draw his distance-time graph.

His parents left home at 13.00 and travelled by car to the same place at 60 km/h. Add their distance-time graph to your diagram and state when and where they overtook their son.

10. A motorist joined a motorway at Junction 5 at 14.30 and travelled northwards at a steady speed of 120 km/h until 15.40 when he turned into a service area. He stayed for 30 minutes and then continued his journey at 105 km/h until he turned off the motorway at 17.20. At 15.00 a lorry joined the motorway at Junction 7, which is 70 km north of Junction 5, and travelled northwards at 80 km/h. Draw the two distance-time graphs and from them find when and where the motorist overtook the lorry.

11. To get from his house to a bus stop a boy had to go 400 metres along his road and then 200 metres along a main road. When he had walked 200 metres he saw a bus pass the end of his road and he started to run. On reaching the main road he found that he had missed the bus. He walked on to the bus stop where he had to wait a few minutes for the next bus. Sketch his distance-time graph.

12. Here is a time-table for two trains.

Station	*Distance* (km)	*Slow train*	*Express*
A	0	dep 10.00	dep 10.20
B	10	arr 10.12	
		dep 10.15	
C	18	arr 10.25	arr 10.30
		dep 10.28	dep 10.35
D	30	arr 10.43	
		dep 10.45	
E	45	arr 11.05	arr 10.50

Draw the distance-time graphs and state when and where the fast train overtakes the slow train. (Assume that each train travels at constant speed between stations.)

So far we have considered only constant (steady) speeds. In practice it takes time to reach a given speed from rest and it takes time to stop. In the above situations these times were very small compared with the times at constant speed and so they were ignored. In the next two questions they are not small.

13. Fig. 6 shows the graph of a goods train which moved from a siding to a signal which was at red.

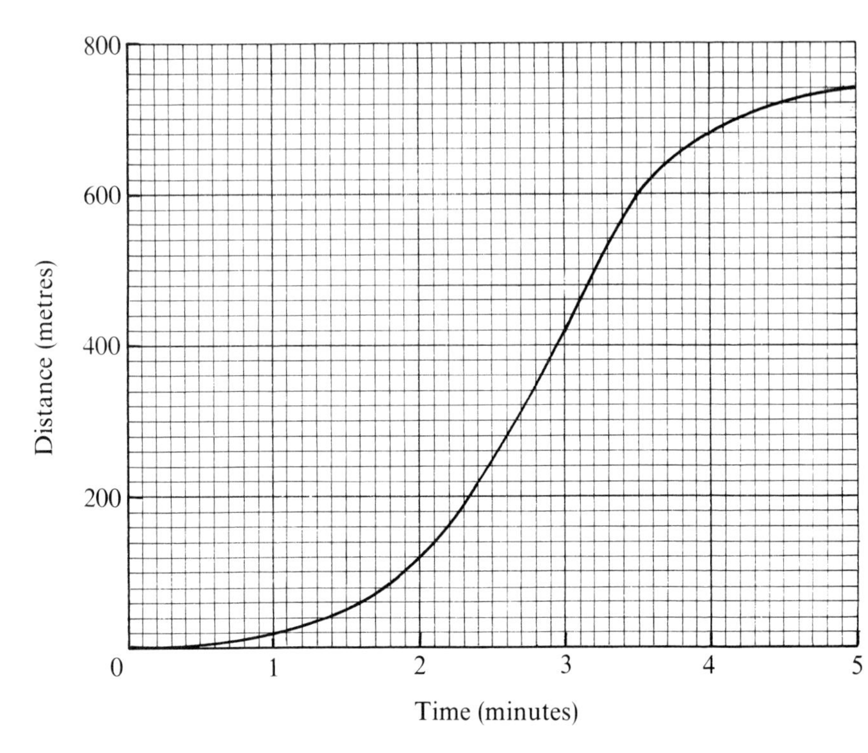

Fig. 6

(i) How far did it travel:
 (a) in the first minute,
 (b) in the first two minutes,
 (c) in the first three minutes?
(ii) State the average speed over each of these three intervals of time.
(iii) State the average speed over the five minutes.
(iv) For how long was the speed increasing?

14. (a) A bus starts from rest, increases its speed gradually, runs at a constant speed, comes to rest at traffic lights, starts again, runs at constant speed and comes to rest at a bus stop. Sketch its graph.

(b) Sketch the graph of a lift which starts at the ground floor (G) and stops at the following levels in turn: 2, 4, G, 3, 5, 2, G.
(c) Sketch the graph of an athlete who runs four times round a track of perimeter 400 m during a 1600 m race. The distance should be labelled 'Distance from the start' and have the numbers from 0 to 400.

SPEED-TIME GRAPHS

In the previous exercise we considered how distance varied with time in certain situations. Now we consider how speed varies with time.

An underground train accelerated from rest to 12 m/s in 15 seconds, stayed at that speed for 20 seconds and then slowed steadily to rest in 10 seconds. Fig. 7 shows the speed-time graph.

When was the speed 8 m/s?

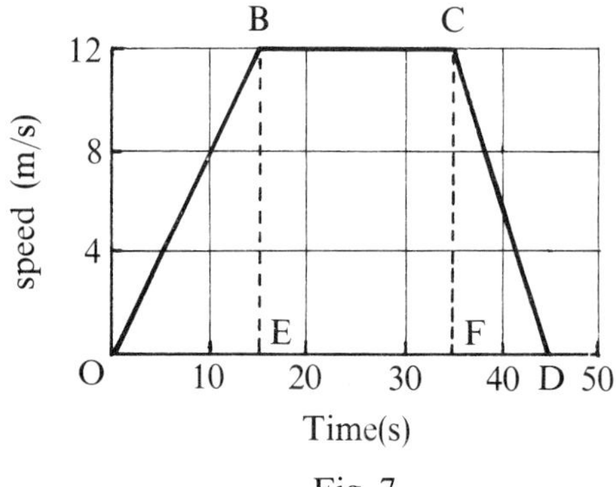

Fig. 7

The speed increased by 12 m/s in 15 s. This is at a rate of $\frac{12}{15}$ m/s each second, that is, 0.8 m/s per second. We say that the *acceleration* was 0.8 m/s per second and write it as 0.8 m/s^2. Notice that the gradient of OB is $\frac{12}{15}$ and so it represents the acceleration.

Over the final 10 seconds the speed decreased from 12 m/s to zero. The *deceleration* or *retardation* was $\frac{12}{10}$ m/s^2, that is, 1.2 m/s^2. This is represented by the gradient of CD.

During the first 15 seconds the average speed was $\frac{1}{2}(0 + 12)$ m/s = 6 m/s and hence the distance travelled was 6 m × 15 = 90 m.

Notice that the area of triangle OEB is calculated in the same way, $\frac{1}{2} \times 15 \times 12$, and so it represents the distance travelled. During the next 20 seconds the train travelled 20×12 m = 240 m which is represented by the area of the rectangle BCFE. Over the final 10 seconds the average speed was 6 m/s and the distance travelled was 10×6 m = 60 m. This is represented by the area of triangle CFD. The total distance travelled was 390 m and is represented by the total area of the quadrilateral OBCD.

Exercise 86

1. (i) A train accelerated steadily from rest to 14 m/s in 20 seconds. What was its acceleration in m/s^2?
(ii) Another train took 20 seconds to accelerate steadily from a speed of 12 m/s to 30 m/s. What was its acceleration?
(iii) A car accelerated from 40 km/h to 100 km/h in 15 seconds. What was its acceleration in km/h per second?

2. A train accelerated steadily from 0 to 60 km/h in 9 minutes, travelled at this speed for 16 minutes, decelerated to 50 km/h in 5 minutes, stayed at this speed for 10 minutes and came steadily to rest in 12 minutes. Draw the speed-time graph. (Suitable scales are 2 cm to 10 minutes and to 10 km/h.)

When was the train travelling at 40 km/h?

State its acceleration and its two decelerations in km/h per min.

3. An electric milk van accelerated steadily from 0 to 8 m/s in 20 s, remained at that speed for 10 s and came to rest in 16 s. Draw its speed-time graph. (Suitable scales are 2 cm to 10 s and 1 cm to 10 m/s).

State its acceleration and deceleration in m/s^2 and calculate how far it travelled.

4. A car accelerated steadily from rest to 90 km/h in 10 s and then to 180 km/h in the next 20 s. It stayed at that speed for 30 s and then decelerated to rest in 30 s. Convert the speed to m/s and draw the speed-time graph. (Suitable scales are 2 cm to 10 s and to 10 m/s.)

(i) When was the car travelling at 40 m/s?

(ii) State its two accelerations and its deceleration in m/s^2.

(iii) Calculate the total distance travelled.

5. (a) Fig. 8 shows a speed-time graph for a lift. Explain it. Remember that if upward speed is taken as *positive*, then downward speed is *negative*.

(b) Sketch the graph for a lift which started at the ground floor, stopped at Floor 1 and Floor 2 and then returned directly to the ground floor.

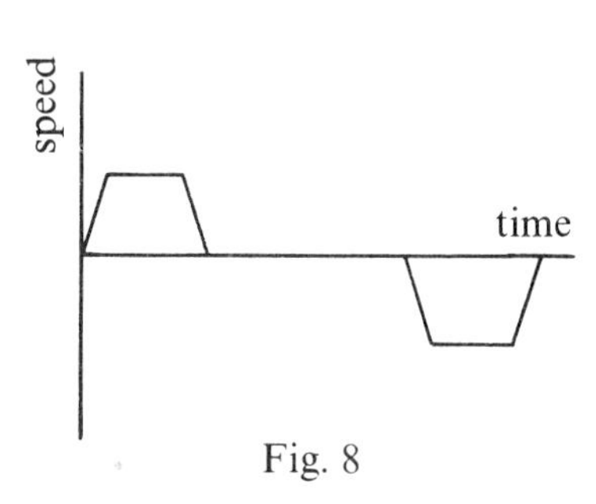

Fig. 8

6. The speedometer readings of a car are shown in the table.

Time from the start (s)	0	5	10	15	20	25	30	35	40
Speed (m/s)	0	6	10	13	14	14	15	18	20

Plot the points and draw a smooth curve through them. (Suitable scales are 2 cm to 5 s and to 5 m/s.)

Estimate:

(i) When the speedometer read 12 m/s.

(ii) The distance the car travelled. (Estimate the area by counting the squares. If you use the suggested scales on paper having small squares of side 2 mm then 4 small squares represent 1 m.)

22 · ACCURACY IN CALCULATIONS

All measurements are approximate. 'Ayeford 15 km' on a signpost means that the distance is 15 kilometres to the nearest kilometre. The distance is really between 14.5 km and 15.5 km.

If the mass of a parcel is 360 g to the nearest 10 g, then it is between 355 g and 365 g.

If the time for a journey is $2\frac{1}{2}$ hours to the nearest $\frac{1}{2}$ h, then it is between $2\frac{1}{4}$ h and $2\frac{3}{4}$ h.

Exercise 87

1. Give the following lengths correct to the nearest metre: 6.7 m, 2 m 40 cm, 35.52 m, 6380 cm.

2. Give to the nearest centimetre: 4.8 cm, 163 mm, 5.19 cm, 317 mm.

3. Give to the nearest 100 g: 3620 g, 5872 g, 9.33 kg, 264 g.

4. Give correct to 2 decimal places: 5.843, 3.726, 0.597, 1.014.

5. Give correct to 1 dec. pl.: 35.79, 23.32, 4.76, 11.97.

6. Give correct to 3 significant figures: 5.273, 7243, 9032, 0.072 68.

7. Give correct to 2 sig. fig.: 14.82, 6725, 849, 0.005 27.

8. Between what two limits do the following lie?
(i) 34 km to the nearest km (ii) 26 cm to the nearest cm
(iii) 530 g to the nearest 10 g (iv) £3200 to the nearest £100
(v) 5.6 s to the nearest 0.1 s (vi) 3 hours to the nearest $\frac{1}{2}$ hour.

9. These numbers are correct to 2 dec. pl. Between what limits do they lie? 6.34, 0.86, 0.07, 9.03.

10. These numbers are correct to 2 sig. fig. Between what limits do they lie? 72, 6300, 0.53, 4.6.

11. Give the number 3.141 59 correct to:
(i) 3 sig. fig. (ii) 3 dec. pl. (iii) 1 dec. pl. (iv) 2 sig. fig.

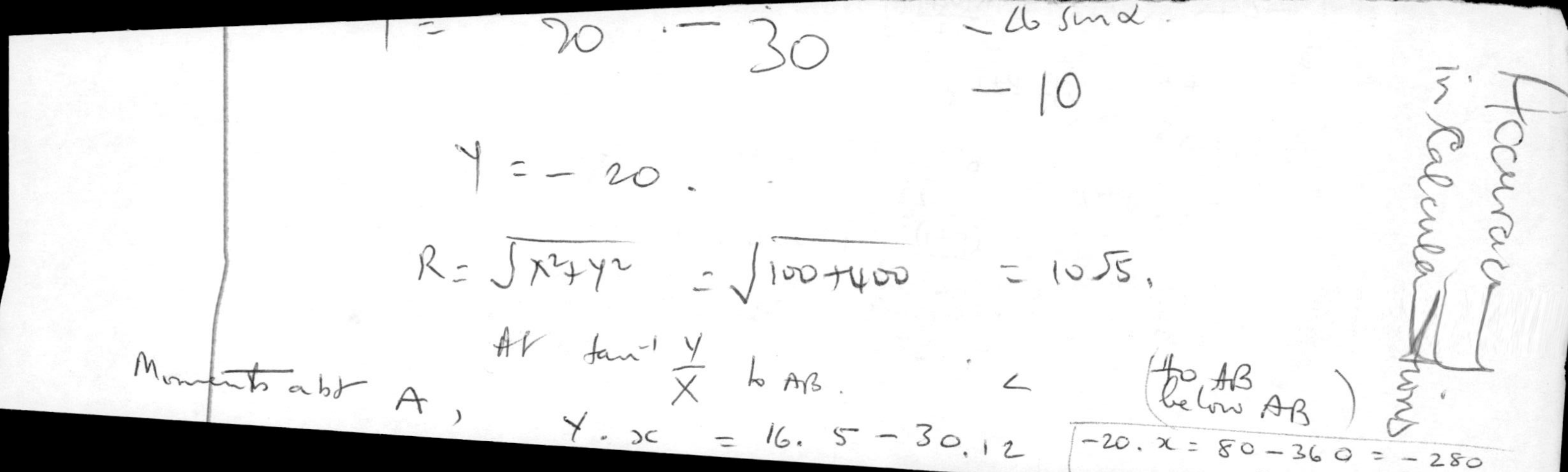

Accuracy in Calculations

$X = 20 - 30 - 26 \sin\alpha$

$- 10$

$Y = -20.$

$R = \sqrt{X^2+Y^2} = \sqrt{100+400} = 10\sqrt{5}.$

At $\tan^{-1}\frac{Y}{X}$ to AB. $\angle$ (to AB below AB)

Moments abt A, $Y \cdot x = 16.5 - 30.12$

$-20 . x = 80 - 360 = -280$

$x = 14.$

$$1^3 + 2^3 + 3^3 + \cdots + n^3 = \left\{\frac{n(n+1)}{2}\right\}^2$$

$$1^3 + \cdots + p^3 = \left\{\frac{p(p+1)}{2}\right\}^2$$

Add $(p+1)^3$

$$1^3 + \cdots + (p+1)^3 = \frac{p^2(p+1)^2}{2^2} + (p+1)^3$$

$$= \frac{(p+1)^2}{2^2}\left\{p^2 + 4(p+1)\right\}$$

$$= \frac{(p+1)^2}{2^2}\left\{p^2+4p+4\right\}$$

$$= \frac{(p+1)^2}{2^2}\left\{(p+2)^2\right\} = \left\{\frac{(p+1)(p+2)}{2}\right\}^2$$

-\$99.

199

12. Explain the difference between 40 cm:
   (i) to the nearest cm
   (ii) to the nearest 10 cm
   (iii) to the nearest $\frac{1}{2}$ cm.

13. Explain the difference between a length given as 52 m to 2 sig. fig. and 52.0 m to 3 sig. fig.

MULTIPLICATION

Mr Thompson has a small rectangular lawn in his garden. To estimate its area, he paces out the length and breadth and finds they are 8 paces and 6 paces. He assumes each pace is 1 metre and calculates the area to be 48 m^2. Suppose that his estimates for the length and breadth are correct to the nearest metre. How accurate is his calculated area of 48 m^2?

The length and breadth can be as large as 8.5 m and 6.5 m which would give the area as 55.25 m^2. They can be as small as 7.5 m and 5.5 m giving an area of 41.25 m^2. Thus Mr Thompson cannot say that the area is 48 m^2. He can only say that it is about 50 m^2 or that it is somewhere between 41 m^2 and 56 m^2.

Exercise 88

1. (i) On graph paper draw a rectangle of length 6 cm and width 4 cm. State its area.
   (ii) Draw a rectangle of length 6.5 cm and width 4.5 cm. Inside it draw a rectangle of length 5.5 cm and width 3.5 cm. State the area of each rectangle.

For each of the following, state the area, assuming that the measurements are exact, and then calculate the greatest and least possible areas assuming that the measurements are correct to the nearest centimetre.

2. A rectangle, length 10 cm, width 8 cm.

3. A rectangle, length 24 cm, width 18 cm.

4. A square of side 9 cm.

5. A triangle, base 12 cm, height 7 cm.

6. A circle of radius 6 cm (to the nearest square centimetre).

7. Calculate the greatest and least possible perimeters for the rectangles of Questions **2** and **3**.

8. (i) The sides of a rectangle are given as 7 cm and 4 cm, correct to 1 sig. fig. The calculated area is 28 cm^2 which is 30 cm^2, correct to 1 sig. fig. Calculate the greatest and least possible areas and state each correct to 1 sig. fig.

(ii) The sides of a rectangle are given as 7.0 cm and 4.0 cm, correct to 2 sig. fig. The area is calculated as 28 cm^2 correct to 2 sig. fig. Calculate the greatest and least possible areas and state each correct to 2 sig. fig. Comment on the result.

9. Betty estimates that she has walked for 3 hours to the nearest $\frac{1}{2}$ hour at a speed of 5 km/h to the nearest 1 km/h. What are the greatest and least possible distances she has walked to the nearest kilometre?

10. An aircraft has an estimated speed of 600 m/s to the nearest 50 m/s. Calculate the greatest and least possible distances it covers in 3 minutes correct to the nearest 10 seconds. Give your answer to the nearest kilometre.

11. A motorist notices that his speedometer shows 54 km/h. Convert this to m/s. He counts up to 140 while passing a lake and believes he is counting seconds. What is his estimate for the length of the lake in kilometres?

He thinks his speedometer may have a 10% error either way. What are the greatest and least possible speeds?

If his estimate of the time may have an error of up to 10 s either way, what are the greatest and least possible lengths of the lake? (2 sig. fig.)

12. Calculate the volume of a cuboid of length 10 cm, width 6 cm and height 4 cm.

If each measurement is correct to the nearest centimetre, what are the greatest and least possible volumes, correct to the nearest cubic centimetre?

PRACTICAL

Find out the length of your walking pace as follows. Walk 10 paces. Measure the distance with a tape measure. Divide the distance by 10.

Find out how many paces you take in one minute. Now estimate your walking speed.

By counting what you think are seconds taken to walk a convenient distance and using your estimated speed, estimate the distance you have walked. Check your estimate by measurement.

DIVISION

A keen hiker, whose watch has stopped, estimates that he has walked 15 km to the nearest kilometre in 3 hours to the nearest half hour. He calculates his speed as 5 km/h. How accurate is this?

The greatest possible speed is obtained by using the greatest possible distance and dividing it by the least possible time. It is $15.5 \div 2\frac{3}{4} = 5.6$ km/h, correct to 2 sig. fig.

The least possible speed is obtained by dividing the least possible distance by the greatest possible time. It is $14.5 \div 3\frac{1}{4} = 4.5$ km/h.

His estimate of 5 km/h could be as much as 0.6 km/h wrong.

Exercise 89

1. A distance is estimated to be between 11 m and 13 m and the time taken is estimated to be between 3 s and 5 s. Speed is found by dividing distance by time. Do the following divisions: $11 \div 5$, $11 \div 3$, $13 \div 5$ and $13 \div 3$. You will see that the greatest speed is obtained by using the greatest distance and least time. What do you use to obtain the least possible speed?

2. Tom estimates that he has walked 9 km to the nearest km in 2 h to the nearest $\frac{1}{2}$ h. Calculate his greatest and least possible speeds to 2 sig. fig.

3. From a map a cyclist estimates that he has travelled 70 km to the nearest 10 km. The time taken was 4 hours by his watch which is correct. Calculate his greatest and least possible speeds.

4. Tom lives 1.8 km from a railway station. He estimates that he walks at 2 m/s. To catch a train at 10.30 he leaves home at 10.13. How many minutes will he have to wait for the train if his estimate is correct?

   If his speed is really $1\frac{1}{2}$ m/s will he miss the train?

5. Calculate to 3 sig. fig. the greatest and least possible values of:
   (i) $8 \div 4$ if each number is given to 1 sig. fig.
   (ii) $8.0 \div 4.0$ if each number is given to 2 sig. fig.

6. Travelling along a motorway in his father's car, Bill estimates, by counting, the number of seconds taken to travel between two telephone boxes which he knows are two kilometres apart. His estimate is 65 s. What speed does this give in:
   (i) metres per second (ii) kilometres per hour? Give your answers to 2 sig. fig.

If his estimate of the time can be as much as 10 s wrong either way, what are the greatest and least possible speeds? Give both m/s and km/h to 2 sig fig.

7. The density of a substance is found by dividing mass by volume. Jane measured the mass and volume of a piece of copper in the Physics laboratory and stated them to be 392 g to the nearest g and 44 cm^3 to the nearest cm^3. What are the greatest and least possible values for the density? Give each answer in g/cm^3 to 2 sig. fig. and compare them with the true density of copper which is 8.9 g/cm^3 to 2 sig. fig.

23 · USING CALCULATORS

Ever since people began to calculate they have tried to develop aids to make calculating easier, quicker and more accurate.

You have probably often used your fingers to help you when counting. The Romans used an abacus. Similar instruments are still used at great speed in the Far East. Early in the 17th Century, Napier devised logarithms, which you used earlier in this book. Later the slide rule was invented. This is based on logarithms. Numbers are multiplied and divided by adding and subtracting lengths which represent logarithms.

The first calculating machine, where the answer could be read off numbered dials was designed by Pascal in 1642. This could only add and subtract. Later Leibnitz designed machines which could multiply. Hand calculating machines using the same principles were used in many schools, colleges and offices in the 1960s.

Electronic computers were developed during the Second World War. In these, moving parts were replaced by radio valves and such computers could work millions of times faster than mechanical calculators, but each needed a large room. With the development of transistors computers became much smaller.

The need for small calculators for space travel led to the development of pocket electronic calculators. Some of these can give you all the information contained in your book of tables, such as square roots, logarithms, sines, cosines and tangents and some can do many other calculations. Some have 'memories' and some can be programmed to do complicated calculations with any numbers they are given.

This chapter contains work which can be done on pocket calculators. You will find that basic adding, subtracting, multiplying and dividing are straightforward. If your calculator does other operations you should study the handbook which will give you instructions.

Exercise 90

1. Give each answer correct to 2 significant figures:

(i) 8.4×7.3 (ii) 0.96×3.7 (iii) 29×44
(iv) $9.6 \div 3.8$ (v) $17 \div 4.6$ (vi) $0.077 \div 0.29$.

2. Change the following fractions into decimals, giving your answers to 3 sig. fig.:

(i) $\frac{14}{17}$ (ii) $\frac{8}{13}$ (iii) $\frac{53}{59}$ (iv) $\frac{36}{97}$ (v) $\frac{16}{75}$ (vi) $\frac{67}{809}$.

3. Try $5 \div 7 \times 7$. Do you get 4.999 99 . . . or 5? The correct answer is obviously 5 but some calculators do give 4.999 99 . . .. Remember that this can happen. If your calculator shows 27.999 999 9 at the end of a calculation, the answer should almost certainly be 28. Try also $11 \div 19 \times 19$.

4. Work out the following. Give each answer to 3 sig. fig.:

(i) $7.2 \times 3.5 \times 9.4$ (ii) $0.82 \times 4.6 \times 1.9$
(iii) 6.3^3 (iv) 0.88^4
(v) $4.8 \times 2.9 \div 1.7$ (vi) $23.7 \times 86.4 \div 427$.

5. Work out the following giving your answers to 3 sig. fig.:

(i) $(24.5 + 17.8) \times 6.4$ (ii) $(92.7 - 81.6) \times 4.4$
(iii) $(7.24 - 2.58) \div 9.62$ (iv) $(23.4 + 72.3) \div 8.46$.

6. A cine camera can be bought by making a down payment of £38.60 and then 24 monthly payments of £7.44. Calculate the total for the 24 monthly payments and then add on the down payment of £38.60.

7. A car was bought by a down payment of £200 and 36 monthly payments of £25.95. Calculate the total cost.

8. A man obtained life insurance by paying £1.71 per week. How much did he pay each year? (52 weeks)

9. A farmer sold 47 tonnes of potatoes at 16p per kilogramme. How much did he get?

10. There are three possible results for a football match: the home team can win, lose or draw. The number of ways of forecasting the results of 7 matches is 3^7. Calculate this number. Calculate also the number of ways of forecasting the results of 10 matches.

11. Eight football teams enter a competition. The names of the teams are placed in a box and drawn out one at a time to decide which teams play each other in the first round. The number of different ways the names can be drawn out is $8 \times 7 \times 6 \times 5 \times 4 \times 3 \times 2$. Work this out. Can you obtain the number of ways for 12 teams on your calculator?

12. (a) Do the following questions by long multiplication. Do as many as you can in 5 minutes. If you do them all in less than 5 minutes, make a note of the time you take:

(i) 57×33 (ii) 79×48 (iii) 0.094×0.82
(iv) 49.6×3.8 (v) 6.83×3.9 (vi) 473×0.68.

(b) Now do the same questions using logarithms. Again do as many as you can in 5 minutes and note the time taken if you do them all.

(c) Do the same questions using a calculator and note the time taken.

SOME NUMBER PATTERNS

Exercise 91

You may find that you cannot do some of these questions because your calculator does not show sufficient figures.

1. Workout: $15873 \times 7, 15873 \times 14, 15873 \times 21$ and 15873×28. How would you obtain a row of fives? Try your idea. How would you obtain a row of sixes?

2. Write down the answers for: $11^2, 111^2, 1111^2$ and so on as far as you can go on your calculator. Study the pattern. If you could continue, would the pattern ever stop?

3. Write down the answers to: $11^2, 11^3, 11^4, 11^5$ and 11^6. Which give symmetric numbers?

4. Write down the answers to: 9×6, 99×66, 999×666 and so on as far as you can. Would this pattern ever stop? On paper work out $(1\,000\,000 \times 666\,666) - 666\,666$. This is $999\,999 \times 666\,666$. Have you still the same pattern?

5. Work out $12\,345\,679 \times 27$ and $12\,345\,679 \times 36$. How would you get a row of fives? How a row of sevens?

6. Without a calculator, work out $5^2 + 5 + 6$ and $7^2 + 7 + 8$. What do you notice about the answers?

$34^2 + 34 + 35 = n^2$. What do you think is the value of n? Try it on a calculator. What is 40^2? Without the calculator, use the above idea to find 41^2.

7. Work out $38^2 - 38 - 39$ and $38^2 - 38 - 37$. Also work out 37^2 and 39^2. Copy and complete $38^2 - 38 - \ldots = \ldots^2$. Now copy and complete: $50^2 - \ldots - \ldots = 49^2$. Without the calculator, use this statement to work out the value of 49^2.

8. Work out $2 \div 11$. You get 0.181 818 1 . . .. We write this recurring decimal as $0.\dot{1}\dot{8}$. Check that $\frac{5}{11} = 0.\dot{4}\dot{5}$. Express $\frac{1}{11}, \frac{3}{11}, \frac{4}{11}, \frac{6}{11}, \frac{7}{11}, \frac{8}{11}, \frac{9}{11}, \frac{10}{11}$ as recurring decimals. What do you notice about the two figures in each case?

9. Express $\frac{1}{3}, \frac{2}{3}, \frac{1}{9}, \frac{2}{9}, \frac{4}{9}, \frac{5}{9}, \frac{7}{9}, \frac{8}{9}$ as recurring decimals.

10. Check that $\frac{1}{7} = 0.142\,857\,142\ldots$. We write this as $0.\dot{1}42\,85\dot{7}$. Express $\frac{6}{7}$ as a recurring decimal. Compare it with that for $\frac{1}{7}$. What do you notice?

Express $\frac{2}{7}, \frac{3}{7}, \frac{4}{7}, \frac{5}{7}$ as recurring decimals. Can you find pairs which have the same pattern as for $\frac{1}{7}$ and $\frac{6}{7}$? Write down the decimal for $\frac{3}{7}$ under that for $\frac{1}{7}$. What do you notice? Continue the column using the decimals for the other sevenths so that the pattern is continued.

11. Express $\frac{1}{13}, \frac{2}{13}, \frac{3}{13}$ etc as recurring decimals. Can you find any patterns like the ones in Question **10**?

12. You may have seen the *Fibonacci* numbers 1, 1, 2, 3, 5, 8, 13, ... Each number is formed by adding together the previous two numbers. Thus $8 + 13$ gives 21 as the next number. Work out the numbers as far as 10 946.

Now express $\frac{3}{2}, \frac{8}{5}, \frac{21}{13}$, etc as decimals and write your answers in a column. Also express $\frac{5}{3}, \frac{13}{8}, \frac{34}{21}$, etc, as decimals and write your answers in a second column. You will notice that the numbers in the first column get larger and those in the second column get smaller. Both get nearer and nearer to a certain number. This is called the *Golden Number*. It is like π, $\sqrt{2}$, $\sqrt{3}$, etc. It cannot be expressed exactly as a fraction or decimal.

If you have $\sqrt{x}$ on your calculator, work out $(1 + \sqrt{5}) \div 2$. It gives the Golden Number to as many decimal places as are possible on your calculator.

13. Start with 3 and 4 instead of 1 and 1 and build up a series like the Fibonacci series. Work out $\frac{4}{3}, \frac{11}{7}$, etc, and also $\frac{7}{4}, \frac{18}{11}$, etc.

DOING WITHOUT CALCULATORS

Calculators should not be used for very simple calculations such as 52×30, $72 \div 10$ and $305 \div 5$. You should be able to do these quicker in your head.

FINDING A REMAINDER

Sometimes we need to do a division sum where the answer is not exact and we need to know the remainder. A calculator does not give us this directly.

Consider $59 \div 8$. The answer is 7 and there is a remainder of 3 because $59 = 8 \times 7 + 3$. Now $\frac{59}{8} = \frac{56 + 3}{8} = 7 + \frac{3}{8} = 7.375$.

If 7 is subtracted from $7\frac{3}{8}$ or 7.375 we get $\frac{3}{8}$ or 0.375. If this is multiplied by 8 we get 3, the remainder.

Again, when 96 is divided by 7 we get 13.7142..., which is $13\frac{5}{7}$. Subtracting 13 we have 0.7142.... Multiplying this by 7 we have the remainder, 5.

Exercise 92

In Questions **1** to **4**, do as many parts as you can without your calculator.

1. Change these fractions into decimals: $\frac{9}{10}$, $\frac{7}{20}$, $\frac{3}{5}$, $\frac{2}{7}$, $\frac{1}{8}$, $\frac{3}{16}$, $\frac{13}{17}$, $\frac{18}{36}$, $\frac{32}{43}$, $\frac{19}{50}$.

2. 43×20, 44×19, 50×66, 53×67.

3. $85 \div 10$, $94 \div 20$, $73 \div 18$, $75 \div 25$.

4. 6.3×0.2, 7.8×0.19, 0.45×0.4, 0.8×0.5.

5. Find the remainders in the following: $276 \div 13$, $578 \div 37$, $9864 \div 59$.

6. I have 24632 Italian lire which I wish to change into £1 notes. The exchange rate is 1517 lire to £1. The exchange bureau does not give any coins. How many £1 notes do I get and how many lire are left over?

7. 723 supporters of the Startown Club are to be taken to an away match in coaches which each hold 46 people. How many coaches are needed and how many empty seats will there be?

8. (a) Express 4.729 hours in hours and minutes, correct to the nearest minute. (Subtract 4 and multiply the decimal, 0.729, by 60.)

(b) Express 8.367 minutes in minutes and seconds correct to the nearest second.

(c) An aircraft has a speed of 880 km/h. How long does it take for a journey of 2465 km? Give your answer in hours and minutes, correct to the nearest minute.

9. Some calculators have tangents, sines and cosines. To find the angle which has a sine of 0.5432 you set this number and press arc sin. This gives 32.901 74 ... which is the angle in degrees. Suppose we want the angle in degrees and minutes. (Remember that $1^\circ = 60$ minutes.) We subtract 32 and turn the decimal of a degree into minutes by multiplying by 60. Do this. You should get $54'$ to the nearest minute and so the angle is $32^\circ\ 54'$.

Express $57.438\,63^\circ$ and $73.845\,72^\circ$ in degrees and minutes.

Find in degrees and minutes the angle which has a tangent of:

(i) 0.6392 (ii) 2.51786 (iii) 4.39758.

AVOIDING MISTAKES

A calculator does not make mistakes but the user can. You must think carefully what you wish the calculator to do.

Suppose you have to work out the total cost of 18 metres of cloth at 67p per metre and 17 metres at 53p per metre. If you press the keys thus $18 \times 67 + 17 \times 53 =$ you will get 64 819 and give your answer as £648.19 which is clearly wrong.

It is wrong because the calculator carries out the working step by step in the order the keys have been pressed. It works out 18×67, adds 17 and then multiplies the result by 53. In fact it does this: $((18 \times 67) + 17) \times 53$. If you watch the window you will see the numbers 18, 67, 1206, 17, 1223, 53, 64 819 appear in turn.

If you have brackets on your calculator you can work this question thus: $18 \times 67 + (17 \times 53) =$

If you have a memory (store) you can work thus:

$$18 \times 67 = \text{sto } 17 \times 53 = \ + \text{rcl} =$$

If you have neither brackets nor memory you must write down the answer to 18×67 and then work out 17×53 and add the two results together.

Exercise 93

1. $6 + (4 \times 3) = 6 + 12 = 18$ and $(6 + 4) \times 3 = 10 \times 3 = 30$
On your calculator press $6 + 4 \times 3 =$. Do you get 18 or 30?

Work out the values of the expressions in the left and centre columns below without a calculator. Then use your calculator for the expression in the right column and compare your results.

$5 \times (7 + 2)$	$(5 \times 7) + 2$	$5 \times 7 + 2$
$12 - (2 \times 5)$	$(12 - 2) \times 5$	$12 - 2 \times 5$
$8 \times (2 - 10)$	$(8 \times 2) - 10$	$8 \times 2 - 10$
$12 \div (3 \times 5)$	$(12 \div 3) \times 5$	$12 \div 3 \times 5$
$30 \div (3 \div 2)$	$(30 \div 3) \div 2$	$30 \div 3 \div 2$.

2. Find the value of $x + yz$ when $x = 9.3$, $y = 2.7$ and $z = 3.8$.

3. Ted was asked to find the total cost of 14 items at 32p each, 36 at 94p each and 28 at 46p each. He gave the answer as £20 941.04. What did he do wrong? What is the correct answer?

4. Without a calculator work out $1 + 3$, $1 + 3 + 5$, $1 + 3 + 5 + 7$ and $1 + 3 + 5 + 7 + 9$. Notice that each answer is a perfect square. The last answer is the square of 5. How many odd numbers were added together for this answer?

Add together the seven odd numbers 1, 3, 5, 7, 9, 11 and 13. Do you get 7^2?

On your calculator press the keys $1 + 3 + 5 + 7 + 9 + 11 + 13 +$. After each + you will see a square number appear. After the last + you should see 49.

Continue with the sequence of odd numbers and + signs. You will not recognise all the squares, but you should recognise certain ones such as 400, 900 and 1600. If these do not appear you have made a mistake.

5. Press the keys as follows $1 + 2 + 3 + 4 + 5 +$. When you press the last + sign, the 5 changes to 15 which is the sum of the integers from 1 to 5. Continue with $6 + 7 +$, etc. At some stage you will see the symmetric number 171 which is the sum of the integers up to 18.

Continuing again you will get a symmetric number just under 600. What is it? What was the last number you added on?

If you continue further you will find that the next symmetric number is over 3000. What is it?

The formula for the sum of the integers $1, 2, 3, 4, \ldots, n$ is $n(n + 1) \div 2$. Test this with $n = 18$, with $n = 34$ and $n = 77$.

6. If you have an x^2 key on your calculator, try $1^2 + 3^2 + 5^2 + 7^2 +$ and so on. At some stage you should have the symmetric number 969. What was the last number you squared? Continuing you soon get a four digit symmetric number. What is it?

7. Form the powers of 2, which are $2^2, 2^3, 2^4, 2^5$, etc, by multiplying repeatedly by $2(2 \times 2 \times 2 \times 2 \times \ldots)$. Count the number of times you press the 2 key. Go on until your display window is full.

Repeat for powers of 3, powers of 4, powers of 5 and so on up to powers of 10. Record your results in a table to show the highest power of each number which your calculator will record. (Some calculators go automatically into standard form when a number is too large for the display window. If you have one of these, stop when this happens and record the last power reached before it happened.)

PRACTICAL ARITHMETIC

The calculations which arise in 'real life' situations usually do not work out easily. A calculator saves time and makes errors less likely.

EXAMPLE 1: *A building society offers 8.75% interest. What interest is received on £364.68?*

$$8.75\% = 0.0875$$

The interest is £364.84 × 0.0875 = £31.9095 which is £31.91 to the nearest penny.

EXAMPLE 2: *When the exchange rate is 118.15 Spanish pesetas to £1, how much British money can be obtained for 5230 pesetas?*

118.15 pesetas = £1

1 peseta = £1 ÷ 118.15

5230 pesetas = (£1 ÷ 118.15) × 5230

= £44.27 to the nearest penny.

Exercise 94

1. A building society pays 8.75% interest. How much interest is received on: (i) £547 (ii) £2089.47.

2. A unit trust pays a dividend of 0.768p per unit. How much is received by a man who owns 3780 units?

3. The cost of hiring a coach to take 37 people on an outing was £24. How much should each person be charged?

4. A householder has a rate demand for £235.41 for a certain year. How much is this per week?

5. When the cost of gas was 15.18p per therm, a householder used 387 therms. What did this cost him?

6. The heights of 9 boys are 152, 148, 156, 165, 143, 159, 168, 150 and 147 cm. Calculate the average height.

7. Calculate the area of a circle of radius 7.32 cm.

8. When £1 = 8.54 French francs:

(i) how much French money is received for £65, and
(ii) how much British money is received for 470 francs?

9. When £1 = 1.7135 US dollars,

(i) how much US money is received for £83, and
(ii) how much British money is received for 324 dollars?

10. At one time electricity bills were worked out as follows: the first 72 units cost 5.689p per unit and the rest cost 2.279p per unit. What was the bill for 1290 units?

11. (i) You may have a square root key on your calculator, but it is interesting to find a square root without using this key. Turn back to page 50 where $\sqrt{29}$ was found to 4 sig. fig.

Do the working on your calculator and continue it until you have $\sqrt{29}$ to as many figures as are possible. Find $\sqrt{88}$ and $\sqrt{456}$ by this method.

(ii) A similar method can be used for cube roots. For $\sqrt[3]{48}$ we need a number x such that $x^3 = 48$. $3^3 = 27$ and $4^3 = 64$. Try 3.5^3, 3.6^3 and 3.7^3. You should find that $3.6 < x < 3.7$. Continue with 3.64^3, etc, until you have $\sqrt[3]{48}$ to as many figures as possible.

12. Here is another method for $\sqrt{29}$:

5 is a rough approximation.

$29 \div 5 = 5.8$ $\quad$ $(5.8 + 5) \div 2 = 5.4$

$29 \div 5.4 = 5.370\ldots$ $\quad$ $(5.370\ldots + 5.4) \div 2 = 5.385\ldots$

Continue in this way until you have as many figures as possible. Use the method for $\sqrt{88}$ starting with 9.

REVISION PAPERS C

REVISION PAPER C1

1. (a) State the values of: 3^2, 3^{-2}, 3^0, 3^{-1}.

(b) Express as powers of 2: 16, $\frac{1}{2}$, 1, $\frac{1}{8}$.

(c) Simplify: $a^{-3} \times a^{-2}$, $b^7 \times b^{-3}$, $c^4 \div c^6$, $d^{-3} \div d^{-2}$.

(d) Write in standard form: 73 000, 0.0084, 4 million, 7 millionths.

2. (a) Find 16% of £8.25.

(b) 40% of a sum of money is £88. Find the sum of money.

3. (a) If an unbiased die is marked with the numbers 1, 1, 2, 3, 5, 6, what is the chance of throwing an even number?

(b) Of 50 parked cars, how many would you expect to have a registration number ending in 7?

4. (a) Calculate $\bar{4}.5 + \bar{2}.9$, $\bar{4}.5 - \bar{2}.9$, $\bar{4}.5 \times 5$.

(b) Use logarithms to calculate:
(i) 0.843×0.726 (ii) $7.26 \div 0.0843$.

5. Solve the equations:

(i) $2^x = 16$ (ii) $10^y = 0.001$ (iii) $n^3 = -27$
(iv) $x^2 - 7x + 10 = 0$ (v) $y^2 + y - 12 = 0$

6. (i) Make a sketch to show the positions of A(2, 3), B(7, 3) and C(4, 9). Join them to form a triangle.

(ii) Mark a point N on AB so that $\widehat{\text{CNA}} = 90°$. State the coordinates of N and the length of CN.

(iii) Find the area of triangle ABC.

(iv) State the tangent of $\widehat{\text{CBN}}$ and use tables to find the size of $\widehat{\text{CBN}}$.

7. (a) Which of the following involve scalar quantities and which involve vector quantities?

(i) The time is 11.00 hours.
(ii) Passing a ball in a soccer match.

(iii) Sighting an aircraft on a radar screen.
(iv) Shooting an arrow.
(b) Which vectors in Fig. 1 are equivalent vectors?

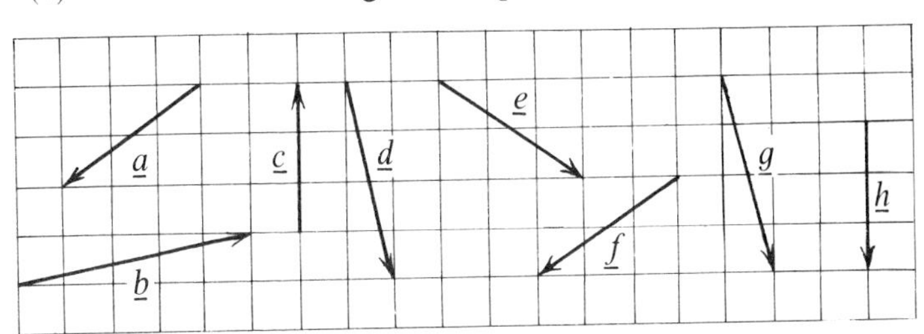

Fig. 1

8. Draw two networks such that one has four nodes of order three and the other has three nodes of order four. (See page 143.)

REVISION PAPER C2

1. (a) Work out the squares of: 0.6, 0.1, 30, $1\frac{1}{4}$.
(b) State the square roots of: 400, 0.04, 0.0036, $2\frac{1}{4}$.
(c) Use tables to obtain the approximate square roots of the following, correct to 3 sig. fig.: 67, 670, 0.67, 0.067.

2. (a) Express as a single fraction:
(i) $\frac{a}{7} + \frac{b}{4}$ (ii) $\frac{3}{c} + \frac{5}{d}$ (iii) $\frac{p}{n^2} + \frac{r}{n}$
(b) Express as two fractions added together:
(i) $\frac{5e + 3f}{15}$ (ii) $\frac{gx + hu}{gh}$ (iii) $\frac{m + 7}{m^2}$.

3. A formula for the ideal mass, m kg, of a man of medium build and height h cm is $m = 0.6\,h - 35$.
(i) Find the ideal mass for a man of height (a) 160 cm (b) 190 cm.
(ii) Make h the subject of the formula.
(iii) A man states that he has the ideal mass of 70 kg. What is his height?

4. (i) For each of the networks in Fig. 2, check that the number of nodes plus the number of regions is two more than the number of arcs.
(ii) Find which of the networks are traversable.

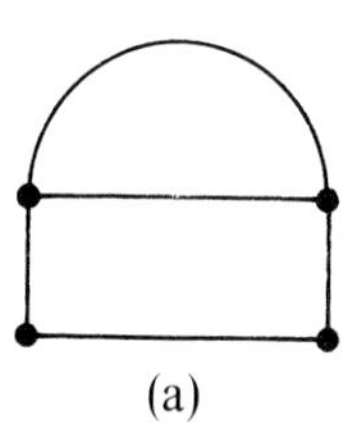

(a)

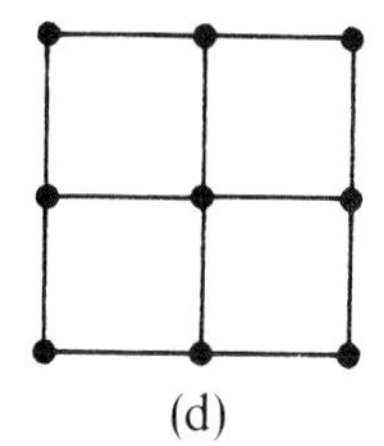

(b)

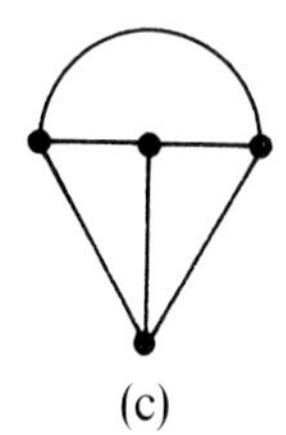

(c)

(d)

Fig. 2

5. Peter went for a walk. His distance-time graph is shown in Fig. 3.

   (i) If he left home at 9.40 a.m., when did he arrive back?
   (ii) How far did he walk altogether?
   (iii) How long did he rest?
   (iv) How fast did he walk on the way home?
   (v) What was his slowest speed?

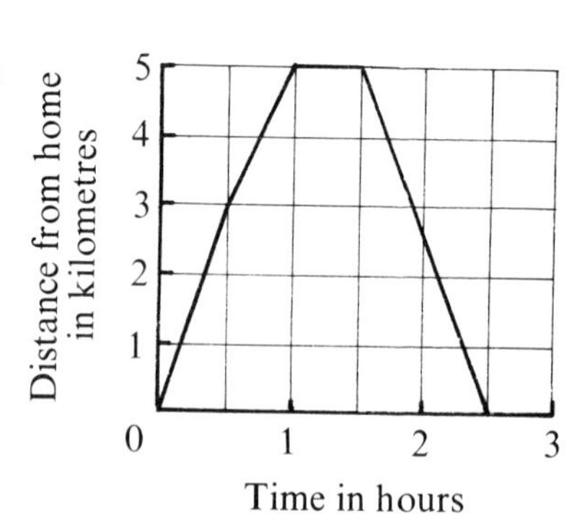

Fig. 3

6. The following statements are wrong. Think what changes should be made to the right-hand side of each and then write down your corrected statement.

   (i) $(a + 5)^2 = a^2 + 25$ (ii) $(b - 8)^2 = b^2 + 16b - 8$
   (iii) $(c + 7)(c - 3) = c^2 - 21$ (iv) $(d - 2)(d - 4) = d^2 - 8d + 6$.

7. In Fig. 4, M is the mid-point of QR. State the size of:

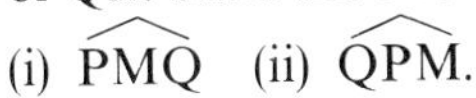

(i) $\widehat{PMQ}$ (ii) $\widehat{QPM}$.

Using the sine or cosine ratio, calculate:

(iii) QM (iv) QR.

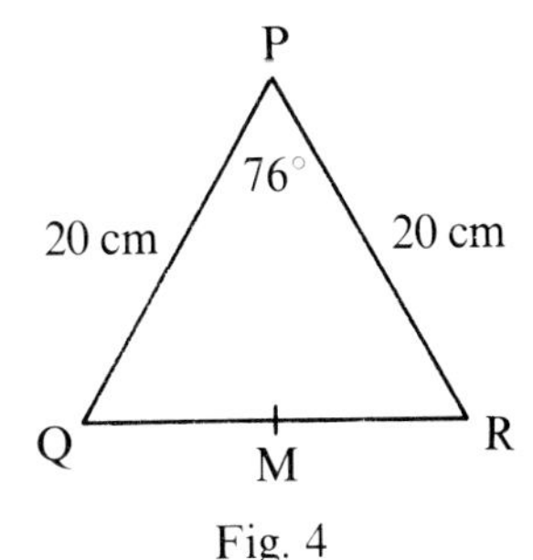

Fig. 4

8. Draw a flow chart for
either (i) washing up after a meal,
or (ii) finding the sum of the squares of the numbers from 1 to 10.

REVISION PAPER C3

1. A car uses 30 litres of petrol when travelling 276 km.

(i) How many litres are likely to be needed for 230 km?
(ii) How far will the car probably go on 35 litres?

Explain why 'likely' and 'probably' have been used.

2. (a) Write as 10^n: 10 000, 0.01, 1 trillion.
(b) State as fractions: 4^{-1}, 3^{-3}, 2^{-4}.
(c) State, in standard form, the answers to: 300×40, $30 \div 500$, and 40^3.

3. Factorise:

(a) (i) $a^2 - 4a$ (ii) $b^2 - 4$
(iii) $c^2 - 5c + 4$ (iv) $d^2 - 3d - 4$.

(b) Solve $x^2 + 4x - 5 = 0$.

4. A ladder of length 4.2 m is placed against the wall of a house and makes an angle of 72° with the ground. How far is the foot of the ladder from the wall and how far up the wall is the top of the ladder?

5. In Fig. 5, M_1 and M_2 are axes of reflection (or mirrors). Triangle A is the object. Copy the figure and on it:
   (i) construct and label as A_1 the image of A in M_1 and
   (ii) construct and label as A_2 the image of A_1 in M_2.

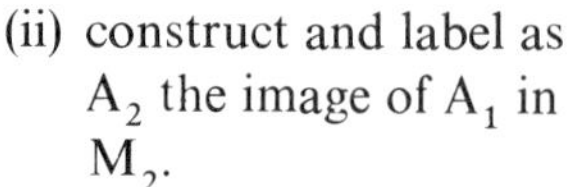

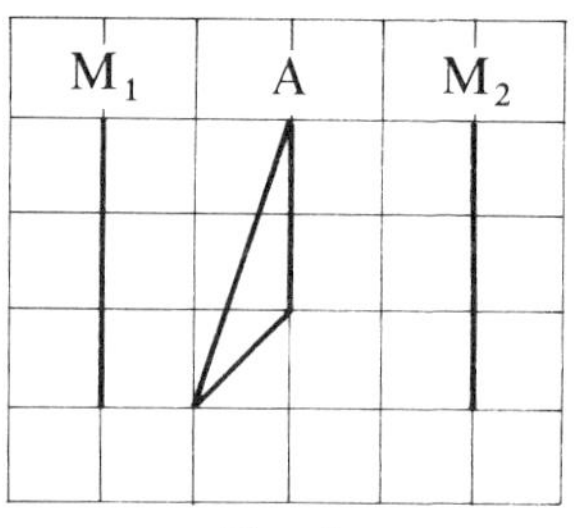

Fig. 5

Show that A_2 is the image of A using a single translation whose distance is twice the distance between the axes M_1 and M_2.

6. (a) Using Fig. 6, write $\overrightarrow{OA}$ and $\overrightarrow{AB}$ as column vectors. State their sum as a column vector.
   (b) On squared paper draw a diagram to represent
   $\begin{pmatrix} 7 \\ 3 \end{pmatrix} + \begin{pmatrix} -2 \\ 3 \end{pmatrix} + \begin{pmatrix} -5 \\ -6 \end{pmatrix}$
   Why do you get a closed figure?

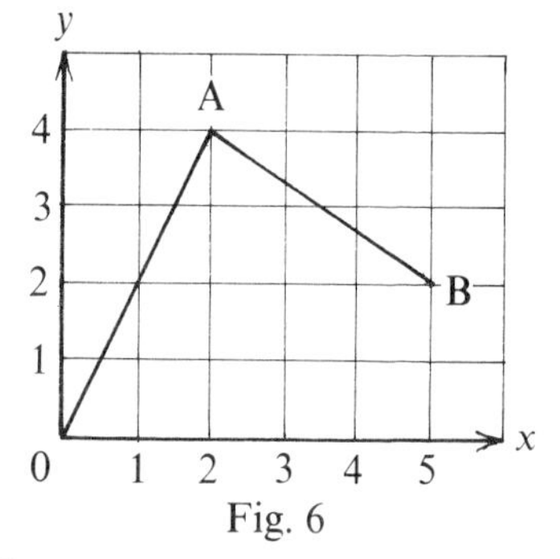

Fig. 6

7. (a) A rectangular plot of land is stated to be 290 m by 140 m, correct to the nearest 10 metres. Between what limits does the area lie?
   (b) Bob estimated that he had walked 14 km to the nearest km in 3 h to the nearest $\frac{1}{2}$ h. Calculate his greatest and least possible average speeds.

8. Use logarithms or a calculator to calculate:
   (i) 723×0.058 (ii) $0.634 \div 0.0983$
   (iii) $(0.873)^3$ (iv) $\sqrt[3]{0.0873}$.

REVISION PAPER C4

1. (a) If $p = -3, q = 2$ and $r = -1$, find the value of:
   $p + q,\ p - q,\ pr,\ q^2 - r^2$ and $p^3 + r^3$.

(b) Write down the factors of $A^2 - B^2$ and use the result to calculate:

(i) $66^2 - 34^2$ (ii) $3.8^2 - 1.8^2$

(iii) $\sqrt{(61^2 - 60^2)}$ (iv) $\sqrt{(5.8^2 - 4.2^2)}$.

2. P and Q are subsets of ξ. Draw suitable diagrams to illustrate the following cases:

(i) $P \subset Q$ (ii) $P \cap Q = \phi$

(iii) $P \cap Q \neq \phi$ (iv) $P \cup Q = \xi$.

Suppose that ξ = {pupils taking a certain Mathematics test}, P = {those who score more than 60%} and Q = {those who score less than 50%}. For each of the cases (i) to (iv) state whether it is true, false or possible.

3. (a) Find the probability that the top card of a shuffled pack is:

(i) the queen of diamonds (ii) a queen (iii) a diamond.

(b) Tom has a chance of 1 in 5 of winning a race and Jim has a chance of 1 in 4 of winning the same race. What is the probability that Tom or Jim wins?

4. The masses of the crew of eight oarsmen and a cox are, in kilogrammes, 76, 84, 77, 88, 89, 86, 90, 74, 56. Find the mean and median masses of the nine men.

Which do you think better represents the data?

What is the mean mass of the eight oarsmen?

5. A temperature of F degrees on the Farenheit scale corresponds to a temperature of C degrees on the Celsius scale where $F = 32 + \frac{9}{5}C$.

(i) Calculate the freezing point ($C = 0$) and the boiling point ($C = 100$) for water on the Farenheit scale.

(ii) Calculate F when $C = 20$ and when $C = -10$.

(iii) Rearrange the formula so that C is the subject.

(iv) Calculate C when $F = 50$, 5 and 140.

6. (a) Write without brackets:

(i) $3a(a + 4)$ (ii) $(b - 7)(b + 5)$ (iii) $(c - 3)^2$.

(b) Solve the equations:

(i) $n^2 - 3n = 0$ (ii) $p^2 - 3p - 10 = 0$ (iii) $t^2 - 25 = 0$.

7. Calculate, correct to 3 sig. fig:

(i) 0.836^2 (ii) $\sqrt{554}$
(iii) $2.33 \times 1.74 \times 6.82$ (iv) $(32.4 \times 59.3) \div 18.4$.

8. A cyclist leaves P at 10.00 and cycles at 15 km/h towards Q which is 25 km from P. At what time does he arrive?

Draw a distance-time graph using 6 cm for 1 hour on the x-axis and 2 cm for 5 km on the y axis.

A car starts from P at 10.40 and travels towards Q at 75 km/h. It stays at Q for 20 minutes and then returns to P at the same speed. Draw its distance–time graph using the same axes.

At what times does the car pass the cyclist and how far from P?

REVISION PAPER C5

1. Solve the following equations:

(i) $\frac{x}{3} - \frac{x}{5} = \frac{4}{3}$ (ii) $\begin{cases} 6x + 2y = 16 \\ 3x + 5y + 2 = 0 \end{cases}$ (iii) $x^2 + x - 12 = 0$.

2. (a) Find, correct to 3 sig. fig., the squares of:
(i) 3.86 (ii) 78.4 (iii) 0.642.

(b) Find, correct to 3 sig. fig., the square roots of:
(i) 59.7 (ii) 423.6 (iii) 0.0752.

3. On squared paper mark the points A (3, 4), B (5,3) and C (4,1). Mark D and E so that the pentagon ABCDE is symmetrical about the line $x = 3$.

(i) State the coordinates of D and E.
(ii) Which two angles of the pentagon are 90°? Explain why.
(iii) Name any equal sides.
(iv) Find the area of the pentagon.

4. (a) Solve the vector equations:

(i) $\begin{pmatrix} x \\ y \end{pmatrix} + \begin{pmatrix} -1 \\ 0 \end{pmatrix} = \begin{pmatrix} 2 \\ 5 \end{pmatrix}$ (ii) $\begin{pmatrix} x \\ y \end{pmatrix} = \begin{pmatrix} 7 \\ 2 \end{pmatrix} + \begin{pmatrix} 3 \\ -1 \end{pmatrix}$

(iii) $2\begin{pmatrix} x \\ y \end{pmatrix} = \begin{pmatrix} 1 \\ -2 \end{pmatrix} + \begin{pmatrix} 5 \\ 4 \end{pmatrix}$.

(b) Given that $\underline{p} = \begin{pmatrix} 3 \\ 0 \end{pmatrix}$, $\underline{q} = \begin{pmatrix} 4 \\ 3 \end{pmatrix}$ and $\underline{r} = \begin{pmatrix} 1 \\ -1 \end{pmatrix}$, draw a diagram to show that $(\underline{p} + \underline{q}) + \underline{r} = \underline{p} + (\underline{q} + \underline{r})$.

5. A car accelerated steadily from rest to 24 m/s in 30 seconds, stayed at this speed for 40 seconds and then slowed steadily to rest in 20 seconds. Draw its speed-time graph using 1 cm for 10 seconds and 1 cm for 5 m/s.

(i) When was the speed 12 m/s?

(ii) What was the acceleration?

(iii) What was the retardation?

(iv) How far did the car travel?

6. Use logarithms or a calculator to calculate, correct to 3 sig. fig: (i) $363 \div 4590$ (ii) $0.459 \div 0.0363$ (iii) $(0.842)^4$ (iv) $\sqrt[4]{0.842}$.

7. Fig. 7 shows a garden roller of radius 28 cm. The drawing bar OH is of length 110 cm. Calculate TH and the angle between OH and the ground. Using $3\frac{1}{7}$ for π, calculate the circumference of the roller. How many times does it turn when pulled a distance of 44 metres?

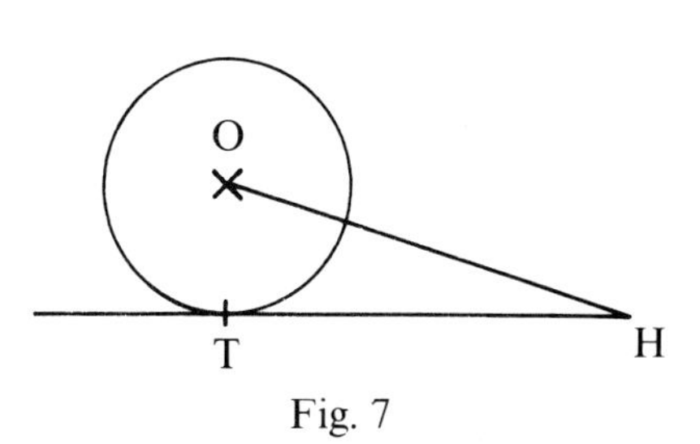

Fig. 7

8. (a) (i) Work through the flow diagram using $x = 26$.

(ii) Work through it again using $x = 50$.

(iii) State the purpose of the flow diagram.

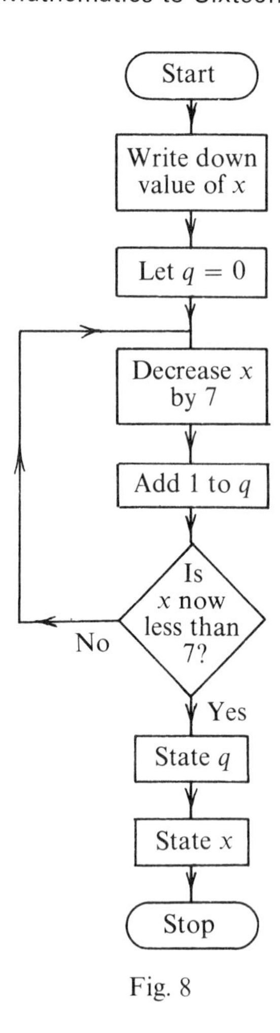

Fig. 8

(b) Draw a flow diagram for the division of one number by another using either logarithms or an electronic calculator.

ANSWERS

Exercise 1 (page 2)

1. $AA_1 = BB_1 = CC_1 = 2$ units
2. Yes; yes
3. (i) A_1B_1 and A_1C_1 (ii) yes (iii) yes
8. P = Q and T = X
9. (i), (iii) and (iv)
11. (i), (a) and (b)

Exercise 2 (page 5)

1. (a), (e) and (f) **4.** $P_1\,(3, -2)$; $P_2\,(-3, 2)$; $P_3\,(-1, 2)$; $P_4\,(3, 2)$
5. $x = 3$; $x = -3$; $x = -1$; $x = 3$ **6.** (1, 0); yes
7. (0, 9) **8.** (i) Yes (ii) yes
11. (0, 2); (0,4); (2, 4); (2, 3); yes **12.** All except (iii)

Exercise 4 (page 14)

4. 21.6 m **5.** 10 cm, 4.5 cm

Exercise 5 (page 17)

1. 32, 64, 128, 256, 512, 1024
2. 9, 27, 81, 243, 729, 2187, 6561
3. 16, 125, 216, 49, 1000, 10000, 100000
4. $3^2, 5^2, 5^3, 3^3, 3^4, 10^5$
5. $2^5, 3^5, 5^6, 7^4, 6^5, 4^8$
6. b^5, c^6, d^4, e^7, f^6
7. $h^{15}, k^{29}, m^{42}, n^{49}, p^{22}$
8. $2^{12}, 3^{15}, 4^{16}, 5^{17}, 6^{19}$
9. $3^3, 5^4, 6^3, 8^3, 9$
10. a^2, b^4, c^6, d, e^6
11. $2^6, 3^7, 4^3, 5^{16}, 10$
12. $f^7, g^3, h, k^{40}, n^{10}$

Exercise 6 (page 18)

1. $3^{10}, 5^9, 6^{12}, 7^4, 10^{12}$
2. $a^5, b^9, c^3, d^{13}, e^{15}$
3. $2^3, 3^3, 5^7, 8, 10^5$
4. f^5, g^3, h^7, k, m^8
5. $a^5, b^2c^3, d^{11}, e^{10}f^2, g^{11}$
6. $2^9, 3^2 \times 5^3, 7^5, 11^5, 2^3 \times 5^4$
7. g^3h^5
8. $\dfrac{p^8}{r^2}$
9. (i) 4^7 (ii) 3^8 (iii) 5^6 (iv) a^8 (v) b^6 (vi) c^9
10. (i) a^2b^7 (ii) c^5d^4 (iii) e^9 (iv) $f^2g^3h^4$ (v) k^6m (vi) np^4
11. (i) a^5b (ii) c^3d^5 (iii) ef^2 (iv) g^6h^3 (v) k^3m^3 (vi) n^4p^2
12. (i) 9 (ii) 8 (iii) 4 (iv) 4
13. (i) 4 (ii) 2 (iii) 5
14. (i) 4 (ii) 8 (iii) 8
15. $b^8, c^{10}, 2^{21}, 7^{20}$

I

Exercise 7 (page 20)

1. $a^6, b^{15}, c^{16}, d^{18}, e^{14}$
2. $2^{12}, 3^{10}, 7^6, 6^{15}, 5^{20}$
3. $\frac{1}{f^4}, \frac{1}{g^7}, \frac{1}{h^3}, 1, \frac{1}{m^8}$
4. $\frac{1}{4}, \frac{1}{27}, \frac{1}{4}, \frac{1}{8}, \frac{1}{100}$
5. 4, 8, 3, 125, 100
6. (ii) $\frac{1}{8}, \frac{1}{9}, 1, \frac{1}{5}$
7. (iii), (iv), (viii), (x)

Exercise 8 (page 21)

1. (i), (iii) **2.** (i) 8×10^2 (ii) 5×10^3 (iii) 7×10^4 (iv) 4×10^6
3. (i) 600 (ii) 30 000 (iii) 9 000 000 (iv) 200 000
4. (i) 3.5×10^2 (ii) 7.6×10^3 (iii) 1.32×10^5 (iv) 6.8×10^4 (v) 9.41×10^2 (vi) 3.625×10^3 (vii) 7.532×10^4 (viii) 8.52×10^7
5. (i) 471 (ii) 683.7 (iii) 5620 (iv) 76000 (v) 9 540 000 (vi) 40 900
6. (i) 5×10^6 (ii) 3×10^{12} (iii) 7×10^{18} (iv) 2×10^{20}
7. (i) 5.98×10^{21} t (ii) 2.28×10^8 km (iii) 8.9×10^3 m (iv) 3×10^5 km/s
8. (i) 10^5 (ii) 10^6 **9.** 12 days **10.** 3.2×10^7
11. 9.5×10^{12} km, 8.1×10^{13} km
12. (i) 6.3×10^5 (ii) 7.29×10^4 (iii) 5.4×10^7 (iv) 2.33×10^4 (v) 9.6×10^4 (vi) 3.65×10^5
13. (i) 2.8×10^4 (ii) 3.6×10^8 (iii) 3×10^6 (iv) 7.2×10^6 (v) 5.04×10^8 (vi) 1.8×10^7
14. (i) 3.6×10^4 (ii) 8×10^6 (iii) 6.27×10^3 (iv) 2×10^4 (v) 4.4×10^6 (vi) 7.3×10^5
15. (i) 1.8×10^3 (ii) 3.2×10^7 (iii) 1.7×10^4 (iv) 5×10^5 (v) 9×10^7 (vi) 1.2×10^2

Exercise 9 (page 24)

1. $10^2, 10^{-2}, 10^4, 10^{-4}, 10^{-2}, 10^0$
2. 1000, $\frac{1}{1000} = 0.001$, 1, $\frac{1}{10} = 0.1$, $\frac{1}{10\,000} = 0.0001$
3. $\frac{1}{9}, \frac{1}{8}, \frac{1}{25}, \frac{1}{6}, \frac{1}{16}$
4. $6^{-3}, 8^{-4}, 4^{-3}, 5^{-3}$
5. $2^1, 2^0, 2^{-1}, 2^{-2}; 2^{-4}, 2^{-5}, 2^{-6}$
6. 27, $\frac{1}{27}$, 1, $\frac{1}{81}, \frac{1}{3}$
7. $3^2, 3^{-2}, 3^{-1}, 3^0, 3^4$
8. $a^{-3}, a^{-5}, a^{-1}, a^0, a^{-7}$
9. $10^5, 10^{-2}, 10^2, 10^{-1}, 10^7, 10^{-2}, 10^3, 10^{-5}, 10^3, 10^{-3}, 10^8, 10^3$
10. $a^{-7}, b^{-5}, c^6, d^{-5}, e^4, f^{-4}, g^3, h^{-3}$

Exercise 10 (page 25)

1. (i) 5×10^{-2} (ii) 8×10^{-4} (iii) 3.9×10^{-1} (iv) 6.4×10^{-2}
2. 4×10^{-2}, 7×10^{-4}, 3×10^{-1}, 2×10^{-3}, 8×10^{-5}, 8.2×10^{-2}, 9.36×10^{-3}, 1.65×10^{-1}, 7.3×10^{-3}, 3.04×10^{-2}
3. 9×10^{-4}, 4.16×10^{-2}, 8.2×10^{-3}, 7.14×10^{-1}, 4.09×10^{-2}
4. 0.0073, 0.000 95, 0.06, 0.000 004 9, 0.0168, 0.57, 0.0806, 0.0005
5. 6×10^{-3}, 7×10^{-6}, 5.5×10^{-5}, 2.3×10^{-3}, 6.05×10^{-3}, 3×10^{-1}
6. (i) 5.46×10^{-5} (ii) 1.58×10^{-19} (iii) 2.4×10^{-5} (iv) 6×10^{-9}
7. (i) 6.4×10^{-3} (ii) 8×10^{-3} (iii) 2.7×10^{-2} (iv) 1.4×10^{-4} (v) 3×10^{-2} (vi) 1.6×10^{3}
8. (i) 3.8×10^{-4} (ii) 2.3×10^{-1} (iii) 4.6×10^{-3}

Exercise 11 (page 29)

1. 4, 1, $\frac{1}{4}$, 0, $\frac{1}{4}$, 1, $2\frac{1}{4}$, 4, $6\frac{1}{4}$, 9; y-axis; 4.84; ± 1.61; 3.24, 7.29, 1.55
2. -27, -15.6, -8, -3.4, -1, -0.1, 0, 0.1, 1, 3.4, 8, 15.6, 27; point symmetry; 5.8, 10.6, 2.71, -2.41
3. -6, -10, -12, -15, 15, 12, 10, 6; symmetry about $y = x$ and about $y = -x$
4. $(-1.5, 2.25)$, $(2, 4)$
5. $(-2, -8)$, $(0, 0)$, $(2, 8)$; $(1.42, 2.9)$
6. 5, -4, -4, 5; 5, -4, -16, -19, -20, -19, -16, -11, -4, 5; translation parallel to y-axis
7. -25, -9, 0, -9, -25; reflection in x-axis

Exercise 12 (page 33)

1. 4 **2.** b **3.** c **4.** 5 **5.** 6 **6.** 6
7. $4h$ **8.** $5k$ **9.** mq **10.** $3u$ **11.** vy **12.** $5w$
13. 12 **14.** $3a$ **15.** bc **16.** d^2 **17.** 28 **18.** fg
19. $6m$ **20.** pt **21.** $3mq$ **22.** $5tu$ **23.** $3wx$ **24.** $4pty$
25. $\frac{2}{7}$ **26.** $\frac{x}{y}$ **27.** $\frac{b}{c}$ **28.** $\frac{d}{e}$ **29.** $\frac{g}{k}$ **30.** $\frac{n}{p}$
31. $\frac{r}{t}$ **32.** $\frac{u}{5}$ **33.** $\frac{1}{3}$ **34.** $\frac{1}{a}$ **35.** $\frac{1}{c}$ **36.** $\frac{1}{d^3}$
37. $\frac{4f}{3}$ **38.** $\frac{h}{g}$ **39.** $\frac{k}{2p}$ **40.** $\frac{4}{3n}$ **41.** $\frac{4}{q}$ **42.** $\frac{3}{x^2}$
43. $\frac{3y}{2v}$ **44.** $\frac{4p}{x}$ **45.** 15 **46.** ab **47.** $5c$ **48.** $4d$
49. 7 **50.** f **51.** g^2 **52.** $3hk$ **53.** 8 **54.** 7
55. p **56.** u^2 **57.** $\frac{1}{8}$ **58.** $\frac{1}{5}$ **59.** $\frac{1}{x}$ **60.** $\frac{1}{y}$

61. $\frac{a}{d}$ **62.** $\frac{(g+h)}{(k+m)}$ **63.** $\frac{2}{3}$ **64.** $\frac{t}{w}$ **65.** $\frac{3}{5}$ **66.** $\frac{4}{3}$

67. $\frac{p}{5}$ **68.** $\frac{3}{x}$ **69.** 6 **70.** 5 **71.** $\frac{(x+y)}{7}$ **72.** $\frac{(a-b)}{3}$

73. $h(m+n)$ **74.** $3(p-q)$ **75.** up **76.** $\frac{3v}{3(a+b)} = \frac{3v}{3a+3b}$

77. $\frac{5(c+d)}{8(c+d)} = \frac{5c+5d}{8c+8d}$ **78.** pm **79.** rt **80.** $u(v+w)$

81. $xk + xn$

Exercise 13 (page 35)

1. $\frac{19}{21}$ **2.** $\frac{7w}{10}$ **3.** $\frac{4x+3y}{12}$ **4.** $\frac{3a}{10}$ **5.** $\frac{py+nx}{xy}$

6. $\frac{5}{6h}$ **7.** $\frac{3}{28m}$ **8.** $\frac{1}{40y}$ **9.** $\frac{2bc+3ad}{6ab}$ **10.** $\frac{12f+10g}{15gf}$

11. $\frac{5}{9}$ **12.** $\frac{5+k}{k^2}$ **13.** $\frac{2m-4}{m^2}$ **14.** $\frac{5p-6}{3p^2}$ **15.** $\frac{13}{30}$

16. $\frac{xw-uy}{uvw}$ **17.** $\frac{25}{7}$ **18.** $\frac{6+f}{3}$ **19.** $\frac{5c+d}{c}$ **20.** $\frac{hk+5}{k}$

21. $\frac{7+pm}{m}$ **22.** $\frac{6q-t}{q}$ **23.** $\frac{v-xy}{x}$ **24.** $\frac{3y^2-3}{y}$ **25.** $3d$, d

26. $5a$, $5b$, 5, 5 **27.** $\frac{x}{5}+\frac{y}{3}$ **28.** $\frac{p}{x}+\frac{t}{y}$ **29.** $\frac{c}{3a}+\frac{d}{2b}$

30. $\frac{5}{k^2}+\frac{1}{k}$ **31.** $\frac{1}{c}-\frac{1}{b}$ **32.** $\frac{h}{3}-\frac{m}{7}$ **33.** $\frac{1}{4}-\frac{1}{n}$ **34.** $\frac{1}{2p}-\frac{1}{4v}$

Exercise 14 (page 36)

1. $\frac{6}{35}$ **2.** $\frac{bc}{xy}$ **3.** $\frac{2b}{3a}$ **4.** $\frac{f^2}{gh}$ **5.** $\frac{3}{4}$ **6.** $\frac{a}{2b}$

7. $\frac{h}{f}$ **8.** $\frac{m}{k}$ **9.** $\frac{5}{7}$ **10.** $\frac{5}{7}$ **11.** $\frac{5}{7}$ **12.** $\frac{2}{3}$

13. 1 **14.** 1 **15.** $\frac{p}{y}$ **16.** $\frac{2}{3}$ **17.** $\frac{1}{3}$ **18.** $\frac{1}{y}$

19. $2a$ **20.** $\frac{2}{bc}$ **21.** $\frac{h}{n}$ **22.** $\frac{5}{2}$ **23.** $\frac{2}{5}$ **24.** $\frac{3w}{v}$

25. $\frac{2}{3}$ **26.** $\frac{3}{2}$ **27.** $\frac{p}{5q}$ **28.** $\frac{a}{12}$ **29.** $\frac{1}{3}$

Exercise 15 (page 38)

1. $\frac{14}{15}$ 2. $\frac{15}{14}$ 3. $\frac{pw}{ru}$ 4. $\frac{ru}{pw}$ 5. $\frac{20}{21}$ 6. $\frac{2}{9}$

7. $\frac{7}{10}$ 8. $\frac{9}{8}$ 9. $\frac{4a}{3b}$ 10. $\frac{hm}{10}$ 11. $\frac{n^2}{6}$ 12. $\frac{7}{p^2}$

13. $\frac{3}{7}$ 14. $\frac{2}{5}$ 15. $\frac{3x}{2y}$ 16. $\frac{kp}{q}$ 17. $\frac{t}{v}$ 18. $\frac{x}{y}$

19. $\frac{3}{5}$ 20. $\frac{2}{3}$ 21. $\frac{3}{2}$ 22. $\frac{9}{4}$ 23. $\frac{q}{2p}$

Exercise 16 (page 39)

1. 12 2. 30 3. 24 4. 10 5. 20 6. 8

7. 40 8. 8 9. 20 10. 2 11. -2 12. $\frac{5}{7}$

13. -7 14. 9 15. $\frac{x}{6}$, 60, 15 16. 30 17. 72

18. $x = 72$, 72°, 48°, 60° 19. 12 m

Exercise 17 (page 42)

1. $y = n + 10$ 2. $y = 90 - x$ 3. (i) 4h (ii) $t = \frac{d}{s}$ (iii) $3\frac{1}{2}$h

4. $y = p - x$; 7.4t 5. $p = k - nx$; 1400 million

6. $q = p + 2xn$ 7. $p = \frac{ny}{100} - x$; £14; loss of £10

Exercise 18 (page 43)

1. $L = \frac{A}{W}$ 2. $p = 4x$ 3. $s = \frac{d}{t}$ 4. $A = \pi dh$

5. $d = \sqrt{13h}$ 6. $V = 4.2r^3$ 7. $a = 180 - 2b$

8. $p = (l + b) \times 2$ 9. $d = \frac{t}{3}$ 10. $d = 4.9t^2$

Exercise 19 (page 44)

1. 4 2. 1.8 3. 45 4. 14 5. ± 3 6. ± 5

7. 81 8. 36 9. $\frac{5}{9}$ 10. 7.5 11. $5\frac{1}{4}$ 12. 52.5

13. 4.5 **14.** 35 **15.** $b - a$ **16.** $\frac{d}{c}$ **17.** fg **18.** $h + k$

19. $\sqrt{n}$ **20.** $\sqrt{\frac{q}{p}}$ **21.** p^2 **22.** $\frac{y^2}{v^2}$ **23.** $\frac{r}{t}$ **24.** $\frac{wy}{u}$

25. $\frac{ab}{c}$ **26.** $\frac{df}{e}$ **27.** $\frac{k - h}{g}$ **28.** $(p + k)n$ **29.** $\frac{c}{\pi}$ **30.** $\frac{V}{lb}$

31. $180 - a - c$ **32.** $\frac{2A}{b}$ **33.** st **34.** $\frac{d}{s}$

35. $\sqrt{\frac{A}{\pi}}$ **36.** $\frac{d^2}{13}$ **37.** $\frac{s}{180} + 2$ **38.** $\sqrt{a^2 - c^2}$

39. $\frac{v - u}{t}$ **40.** $9t^2$

Exercise 20 (page 45)

1. (i) (a) 15 m (b) 20 m (ii) $k = \frac{18}{5}$ m (iii) (a) 90 km/h (b) 153 km/h
2. (i) 60 p (ii) $g = \frac{1}{20}w + 9$ (iii) (a) 13 years (b) 15 years
3. (i) 40 (ii) $h = \frac{x^2}{200}$ (iii) 18
4. $t = \frac{600}{v}$ (i) 8, 5 (ii) $v = \frac{600}{t}$ (iii) 80 km/h
5. (i) $y = 180 - 2x$ (ii) 56 (iii) $x = 90 - \frac{1}{2}y$ (iv) 37
6. (i) $v = 16x^2$ (ii) $x = \frac{\sqrt{v}}{4}$ (iii) 5
7. (i) 108°, 144° (ii) $n = \frac{360}{180 - x}$ (iii) (a) 9 (b) 18 (c) 15
(iv) $n = 5\frac{1}{7}$. A regular polygon cannot have an angle of 110°
8. (i) Each takes 6 counters (ii) 128 (iii) 33 (iv) $p = \frac{200 - n}{6}$ (v) 30
9. (i) Each costs x p (ii) 12 (iii) $n = \frac{100C}{x}$ (iv) 20
10. (i) £360 is basic cost. £10 needed for each guest (ii) 520
(iii) $n = \frac{C - 360}{10}$ (iv) 22 (v) £$30n$, $n = 18, 19$
11. (i) $h = \sqrt{(100 - d^2)}$ (ii) (a) 8 (b) 9.54

Exercise 21 (page 48)

1. (i) 49 (ii) 4900 (iii) 256 (iv) 529 (v) 5.29 (vi) 52 900
(vii) 33.64 (viii) 17.64

2. (i) 0.09 (ii) 0.81 (ii) 0.0081 (iv) 0.0001 (v) 0.0144 (vi) 1.44 (vii) 14 400 (viii) 1 440 000
3. (i) 6 (ii) 30 (iii) 0.6 (iv) 0.3 (v) 40 (vi) 1.2 (vii) 20 (viii) 0.2
4. (i) 10 m (ii) 50 m
5. 44.89, 6.7
6. 28.09, 28.09
7. 115 600, 115 600
8. $\frac{4}{9}, \frac{1}{16}, \frac{9}{49}, 2\frac{1}{4}, 12\frac{1}{4}$
9. $\frac{3}{4}, \frac{1}{3}, 1\frac{1}{2}, 1\frac{1}{3}, 2\frac{1}{2}$
10. 41 cm^2
11. 12.6
12. 548

Exercise 22 (page 50)

1. 7.453, 2.434, 39.31, 53.88
2. 12.25, 24.01, 70.56, 33.64
3. 2.170, 6.462, 19.67, 33.34
4. 8.762, 876.2, 0.087 62
5. 2.341, 23 410, 0.000 234 1
6. 70.39, 0.7039, 7039
7. 57.46, 574 600, 0.5746
8. 12.41, 1241, 124 100
9. 36.86, 0.3686, 0.003 686
10. 69.80, 6980, 0.006 980

Exercise 23 (page 52)

1. 6.633, 8.307, 2.098, 2.627
2. 1.606, 2.640, 3.688, 9.160
3. 2.007, 2.750, 4.868, 8.289
4. 1.877, 2.238, 7.902, 6.564
5. 16.22, 29.92, 79.50, 87.46
6. 56.90, 23.83, 90.63, 98.57
7. 3, 9.487, 30, 94.87
8. 1.581, 5, 15.81, 50
9. 0.3742, 0.8246, 0.7071, 0.8367
10. 0.2522, 0.1253, 0.9338, 0.9654
11. 0.5320, 0.1682, 0.053 20, 0.016 82
12. 0.2634, 0.8331, 0.008 331, 0.026 34
13. 21.75, 6.875, 2.175, 0.6875
14. 9.240, 29.22, 0.2922, 0.9240
15. (i) 3.5 (ii) 1.1 (iii) 6.7 (iv) 16 (v) 0.62
16. 7.6 cm
17. 11
18. 7.6
19. 7.1
20. 23
21. 5.2 m
22. 16 cm
23. 7.6, 9.8, 6.7
24. 8.1 m

Exercise 24 (page 54)

1. (i), (iv), (vi), (vii)
2. 150 m
3. 27 cm, 36 cm
4. (i) 13 cm (ii) 3:1 (iii) 36 cm, 39 cm
5. (ii) 4 cm, 7.5 cm (iii) 60 cm, 68 cm

7. 30 **8.** 25, 5, 20
9. (i) 4, 3, 5 (ii) 6, 8, 10 (iii) 8, 15, 17
10. (i) 20, 21, 29 (ii) 56, 33, 65

Exercise 25 (page 59)

1. (i) {1, 2, 3, 4, 5, 6} (ii) {head, tail} (iii) {0, 1, 2, 3, 4, 5, 6, 7, 8, 9} (iv) {spades, clubs, hearts, diamonds}
2. (i) p(2) = $\frac{1}{6}$ (ii) p(head) = $\frac{1}{2}$ (iii) p(vowel) = $\frac{5}{26}$ (iv) p(ace) = $\frac{1}{13}$
3. (i) $\frac{4}{9}$ (ii) $\frac{5}{9}$ (iii) $\frac{8}{9}$ (iv) $\frac{1}{3}$. Equally likely outcomes
4. (i) $\frac{2}{3}$ (ii) $\frac{1}{3}$ (iii) 0 (iv) 1
5. $\frac{3}{13}$, $\frac{10}{13}$

Exercise 26 (page 62)

1. $\frac{8}{15}$ **3.** (i) $\frac{1}{2}$ (ii) 1 (iii) 0
4. (i) $\frac{6}{11}$ (ii) $\frac{5}{11}$ (iii) 1 (iv) 0 (v) $\frac{3}{11}$ (vi) $\frac{10}{11}$
5. $\frac{1}{50}$ **6.** (i) $\frac{4}{11}$ (ii) $\frac{7}{11}$ (iii) $\frac{3}{11}$
7. (i) $\frac{1}{2}$ (ii) $\frac{3}{8}$ (iii) $\frac{1}{8}$ (iv) $\frac{5}{8}$ (v) $\frac{5}{8}$ (vi) 0
8. (i) $\frac{1}{2}$ (ii) $\frac{1}{4}$ (iii) $\frac{1}{13}$ (iv) $\frac{3}{4}$ (v) $\frac{4}{13}$ (vi) $\frac{3}{13}$
9. (i) $\frac{1}{5}$ (ii) $\frac{2}{5}$ (iii) $\frac{4}{5}$ (iv) $\frac{3}{5}$
10. (i) $\frac{1}{3}$ (ii) $\frac{2}{3}$ (iii) $\frac{2}{3}$ (iv) 0

Exercise 27 (page 63)

1. (i) $\frac{1}{4}$ (ii) $\frac{3}{4}$ (iii) 1 (iv) 0
2. (i) $\frac{1}{3}$ (ii) $\frac{1}{3}$ (iii) 0 (iv) $\frac{2}{3}$
3. (i) $\frac{2}{3}$ (ii) $\frac{1}{3}$ (iii) $\frac{1}{6}$ (iv) $\frac{5}{6}$
4. (i) $\frac{5}{6}$ (ii) $\frac{1}{2}$ (iii) $\frac{1}{2}$
5. (i) $\frac{1}{9}$ (ii) $\frac{8}{9}$ (iii) $\frac{4}{9}$ (iv) $\frac{5}{9}$ (v) $\frac{4}{9}$ (vi) $\frac{5}{9}$
6. (i) $\frac{1}{3}$ (ii) $\frac{1}{6}$ (iii) $\frac{5}{6}$ (iv) $\frac{5}{6}$
7. (i) $\frac{7}{12}$ (ii) $\frac{3}{4}$ (iii) $\frac{3}{4}$

Exercise 28 (page 65)

1. (i) 20 (ii) 60 (iii) 40
2. 40, 60 **3.** 45 **4.** 2 **5.** $\frac{1}{20}$, 40
6. 24 **7.** 219 **8.** 250, 200, 150 **9.** 10

Exercise 29 (page 66)

1. 1, 2, 4, 8, 16, 32
2. 4, 8, 32, 64, 128, 256, 512, 1024, 2048

3. $2^5, 2^7, 2^{12}$ **4.** (i) 2^8 (ii) 2^{12} (iii) 2^8 (iv) 2^{15}
5. (i) 256 (ii) 1024 (iii) 256 (iv) 2048
6. (i) 16 (ii) 128 (iii) 16 (iv) 128

Exercise 30 (page 68)

1. 0.5551 **2.** 0.5557 **3.** 0.8727 **4.** 0.8731 **5.** 0.7545
6. 0.4965 **7.** 0.4530 **8.** 0.2304 **9.** 0.6021 **10.** 0.9031
11. 0.0346 **12.** 0.9495 **13.** 8.903 **14.** 7.467

Exercise 31 (page 68)

1. 1.3711, 2.3711, 3.3711, 4.3711
2. 2.9274, 4.9274, 6.9274, 1.9274
3. 1.0334, 4.0334, 2.0334, 5.0334
4. 2.6990, 3.6990, 1.6990, 4.6990
5. (i) 1 (ii) 4 (iii) 3 (iv) 2 (v) 1 (vi) 0
6. 0.8987, 1.8987, 2.8987, 3.8987
7. 0.7185, 2.7185, 4.7185, 1.7185
8. 0.5416, 3.5416, 1.5416, 5.5416
9. 0.6021, 3.6021, 2.6021, 4.6021
10. 0.9638, 2.9638, 1.9638, 5.9638
11. 0.9218, 2.9218, 5.9218, 6.9218
12. 0.2049, 1.2049, 4.2049, 5.2049

Exercise 32 (page 70)

1. 7.328, 7328, 73.28, 732.8
2. 5.675, 567.5, 56.75, 56 750
3. 2.870, 28.70, 2870, 287 000
4. 1.810, 181.0, 18.10, 1810
5. 9.131, 9131, 91.31, 91 310
6. 1.115, 11.15, 1115, 11 150
7. 1.232, 12.32, 12 320, 123.2
8. 41.21, 41 210, 4.121, 412.1

Exercise 33 (page 71)

1. 31.09 **2.** 402.8 **3.** 334.4 **4.** 532.5 **5.** 22.93
6. 19.89 **7.** 22.61 **8.** 8.489 **9.** 1553 **10.** 106.3
11. 7476 **12.** 205.0 **13.** 9.471 **14.** 147.4 **15.** 51.41
16. 120.8 **17.** 248.2 **18.** 9.281

Exercise 34 (page 71)

1. 3049 **2.** 317.3 **3.** 31.54 **4.** 147 700 **5.** 23.50
6. 3.351 **7.** 5.598 **8.** 2.357 **9.** 17.28 **10.** 1.764
11. 9.237 **12.** 3.752

Exercise 35 (page 73)

1. 36.6 **2.** 48.5 **3.** 267 **4.** 396 **5.** 8.28
6. 108 **7.** 9.15 **8.** 6.20 **9.** 47.4 **10.** 495
11. 17.9 **12.** 50.2 **13.** 383 **14.** 5030 **15.** 2.47
16. 138 **17.** 2.15 **18.** 41.8 **19.** 710 **20.** 305
21. 71.8 **22.** 1130

Exercise 36 (page 73)

1. 3030cm^2 **2.** 405 cm^2 **3.** 6.88 cm **4.** 30.2 cm
5. 951 cm^2 **6.** 8.15 cm **7.** 6520 g **8.** 15.5 s
9. 1 050 000 **10.** 2.06×10^{14}

Revision Paper A1 (page 74)

1. (i) {1, 3, 5, 7, 9} (ii) {2, 3, 4, 6, 8, 9} (iii) {1, 5, 7}
(iv) {1, 2, 4, 5, 7, 8} (v) {1, 5, 7}. $(A \cup B)'$ and $A' \cap B'$ are the same
2. (a) 125, 243, 36, 64 (b) $3^4, 2^5, 5^4, 10^4$ (c) a^9, b^6, c^3, d^{12}
3. (a) 73° (b) 120° (c) 150°
4. (a) $\frac{3}{7}, \frac{3a}{2}, \frac{b}{2}$ (b) (i) 15 (ii) 6
5. (a) $\begin{pmatrix} 10 & -3 \\ -1 & 2 \end{pmatrix}$, impossible, $\begin{pmatrix} 1 \\ -3 \end{pmatrix}$, impossible, $\begin{pmatrix} 16 & -14 \\ 22 & -18 \end{pmatrix}$,
$\begin{pmatrix} 9 & 17 \\ -7 & -11 \end{pmatrix}$
(b) (i) $\begin{pmatrix} 2 & -7 \\ -7 & 4 \end{pmatrix}$ (ii) $\begin{pmatrix} 10 & 13 \\ 16 & -7 \end{pmatrix}$
6. (a) $\frac{7}{10}, \frac{11}{20}, \frac{6}{25}, \frac{3}{8}$ (b) 30%, 35%, 75%, $62\frac{1}{2}$% (c) (i) £1.35 (ii) £1.35
7. (a) (i) Reflection (ii) rotation (iii) translation
(b) (i) (6, 5) (ii) (−2, 11)
8. 5.8, 1.7, 2.6, 2.2, 1.6; (−2.8, 7.8), (1.8, 3.2)

Revision Paper A2 (page 75)

1. (a) £3.40, £5.10 (b) 308 g
2. (a) 6×10^4, 5.4×10^3, 7.8×10^{-3}, 3×10^{-4}

(b) 42 300, 900, 0.006, 0.054
(c) (i) 6×10^6 (ii) 3.5×10^8 (iii) 1.5×10^4 (iv) 4.2×10^{-4}

3. (a) (i) $\frac{4}{9}$ (ii) $25a^6$ (iii) 0.04 (iv) $1\frac{9}{16}$
(b) (i) 0.3 (ii) $\frac{3}{5}$ (iii) $6b^{18}$ (iv) $2\frac{2}{3}$

4. (i) 327 (ii) 20.8 (iii) 62 600

5. 31 200 m^2

6. (a) $7a - 9b$ (b) 0.3 (c) 3

7. (a) 0 (b) 1 (c) $\{1, 2, 3, 4, 5, 6\}$ (i) $\frac{1}{6}$ (ii) $\frac{2}{3}$

Revision Paper A3 (page 77)

1. (i) {p, q, s, u, v} (ii) {p, u} (iii) {q, r, s, t, v, w} (iv) {r, t, w};
(ii) and (v) are true

2. (a) $12a^7, 2b^5, 16c^6, d^8$ (b) 4, -3, 0

3. (a) 3, 2, 0 (b) 455, 8.24

4. (a) $1\frac{4}{15}, \frac{ad + bc}{bd}, \frac{4 - 2f}{f^2}$ (b) 8, -9

5. (i) 27, 45 (ii) $\frac{4s^2}{81}$ (iii) 16, 400

6. (b) (1, 2), (3, 5), (4, 3)

7. 132 cm, 1386 cm^2

8. (a) 25 cm (b) 8.66 cm

Revision Paper A4 (page 78)

1. (a) Each is less than 45° (b) 30°, 80°, 70°

2. (a) 0.23, 0.07, 0.185, 0.045 (b) 40 %, 15 %, 90 %, $3\frac{1}{2}$ %

3. (i) $\frac{9}{25}$ (ii) $\frac{19}{25}$ (iii) $\frac{19}{25}$ (iv) $\frac{2}{5}$

4. (a) $1\frac{1}{3}, \frac{a}{3b}, 3c$ (b) $2\frac{2}{3}$

5. (ii) 135, 75 (iv) 50, 125 (v) 76

6. (a) (i), (iii), (v), (vi) (b) 11, 70

7. 34 cm, 92 cm^2

8. 2.1, 3.5, 1.8; $(2\frac{1}{2}, 6\frac{1}{4})$; (1.3, 4.7)

Revision Paper A5 (page 80)

1. (a) $10^3, 10^{-3}, 10^{-4}, 10^6$ (b) $\frac{1}{9}, \frac{1}{5}, \frac{1}{8}, \frac{1}{16}$ (c) $e^{-8}, f^{-12}, g^{12}, h^{-6}$

2. (i) 20 100, 123, 3.87

3. (a) 15 (b) $x = 2, y = -2$

4. (a) $5(a + 3b), d(c - d), 3f(f - 2g + 3h)$ (b) ± 4, impossible, 2, -3

5. 66 m^2

7. (i) 2 (ii) 4, 5 (iii) $-2, -1, 5$

Exercise 37 (page 84)

1. (i) 8 (ii) 23 (iii) 1.4 (iv) 28.5
2. 65p **3.** 30 **4.** 27, 25 **5.** 16, 13, 15
6. 55 000 **7.** Startown, Startown **8.** 570 **9.** 148
10. 6.25, 6.4, first; 6.9, 7, still first **11.** 5 616 000 tonnes

Exercise 38 (page 86)

1. (i) 5 (ii) 20 (iii) 25 (iv) 0.4
2. 45 kg **3.** Median is 12, mean is 22
4. (i) £11 (ii) 6.5 km (iii) 40 g (iv) 6.5 h
5. Median is 26, mean is 40 **6.** Median is 65p, mean is 80p
7. Median is 4, mean is 6

Exercise 39 (page 87)

1. (i) 21 years (ii) 16 years (iii) 28 years
2. £1000, £4000, £4010 **3.** 47p, 67p **4.** 480
5. £3248 **6.** 28, 24 (i) 40, 36 (ii) 8, 4 (iii) 56, 48

Exercise 40 (page 89)

1. (i) 75 (ii) 260 (iii) 11.7 **2.** Mean is 1980; median is 1990
3. Boys: mean 153 cm, median 151 cm. Girls: mean 145 cm, median 142 cm
4. Mean 7.9 %, median 7.7 %
5. Widows: mean 72 years, median 71 years. Widowers: mean 69 years, median 67 years
6. Boys: mean 56, median 56. Girls: mean 58, median 56
7. Mean 261 h **8.** 12.8 s, 14.6 s; 30 km/h, $26\frac{2}{3}$ km/h

Exercise 41 (page 91)

1. 157.6 cm **2.** 70 years **3.** 47.4 kg **4.** 2.5 cm **5.** 210 h
6. 216, 173.6, 198.4, 114 hours, 5.75 hours
7. (i) 800 kg (ii) 460 kg (iii) 42 kg
8. $\frac{1}{3}(x + 2y)$

Exercise 42 (page 95)

1. one goal **2.** 96 goals **3.** 2 letters, $\frac{1}{3}$
4. 1 to 3 years; almost 43 % **5.** Grade 4, 50 %
6. Frequencies are 40, 80, 100, 120, 100, 100. Modal class is £1500 to £2000
7. 3 days' absence **8.** 1 accident; 450 **9.** 1 to 2 kg

Exercise 43 (page 97)

1. (ii) and (iii) **2.** (i), (v) and (vii) are discrete
4. (i) 47.5 m to 48.5 m (ii) 15.5° to 16.5° (iii) 73.25 kg to 73.75 kg
(iv) 12.3 s to 12.5 s (v) 9.55 cm to 9.65 cm (vi) 725 km to 735 km

Exercise 44 (page 99)

1. 6 h, 46 to 51 h **2.** 31 to 37 marks **3.** 920, 1.15
4. 3 to 5 km **5.** 10 s, 45 to 54 s

Exercise 45 (page 102)

2. 0.84 **3.** 0.47, 0.58, 0.70, 1.19, 1.43, 1.73; 1
4. 0.62 or 0.63, 0.93, 1.60, 1.88, 38°, 47°
5. 2.35 cm, 18.8 cm, 2.82 m, 9.4 m
6. (i) 28.6 cm (ii) 5.60 cm (iii) 6.92 cm
7. (i) 7.5 cm (ii) 39.9 cm
8. 0.49, 26° (i) 1.2, 50° (ii) 0.7, 35° (iii) 1.48, 56°
9. (i) 7 m (ii) 20 m (ii) 10 m

Exercise 46 (page 105)

1. (i) 0.4921 (ii) 0.9457 (iii) 0.0981 (iv) 1.4826 (v) 3.9232 (vi) 0.8352
2. 0.128, 0.2126, 0.5774, 0.7265, 1.7321, 2.7475. No, no, no
3. 5.32 **4.** 2.60 **5.** 5.40 **6.** 3.64 **7.** 39.3
8. 6.89 **9.** 101 **10.** 3.17 **11.** 0.924 **12.** 4.20
13. 4.04 **14.** 8.88 **15.** 14.3 **16.** (i) 3.46 (ii) 2.91
17. 0.18, 0.58, 0.84, 1.19, 1.43, 1.73, 2.14, 2.75, 3.73

Exercise 47 (page 108)

1. 10.4 m **2.** 33.3 m **3.** 56°, 74.1 m **4.** 15.4 cm, 76.9 cm^2
5. 17.2 cm **6.** 11.5 m **7.** 1300 m **8.** 45.2 m
9. 15.1 m **10.** (i) 10.6 m (ii) 25.6 m (iii) 15.0 m
11. 436 m

Exercise 48 (page 111)

1. (i) 17° (ii) 25°42′ (iii) 66°24′ (iv) 33°13′ (v) 72°31′ (vi) 59°38′
2. 0.8, 38°39′ **2.** 0.75, 36°52′ **4.** 1.8, 60°57′
5. 1.5714, 57°32′ **6.** 1.3333, 53°8′ **7.** 2.4, 67°23′
8. 0.7875, 38°13′ **9.** 0.5667, 29°32′ **10.** (i) 31° (ii) 62°
11. 61° **12.** 34° **13.** 66° **14.** 54° **15.** 36°
16. 23° **17.** 78°, 102° **18.** 99° **19.** 45°, 63°, 72°, 76°

Exercise 51 (page 119)

3. Translation of 1 unit to the left and 5 down
5. Translation of 1 unit to the right and 3 down
6. (ii) 9 **7.** 2 **8.** Centre O, scale factor $\frac{4}{3}$

Exercise 52 (page 124)

2. (a) $fg + 4f + 6g + 24$ (b) $(n + 7)(n + 3) = n^2 + 10n + 21$
3. (i) $km + 2k + 3m + 6$ (ii) $np + n + 4p + 4$ (iii) $q^2 + 13q + 40$
(iv) $u^2 + 16u + 63$ (v) $12y^2 + 17y + 6$
4. (i) $3a + 6$ (ii) $bc + 4b$ (iii) $xy + 3x + y + 3$
(iv) $mn + 5m + 7n + 35$ (v) $pr + pq + qr + q^2$
(vi) $4 + k + 4n + kn$ (vii) $h^2 + 8h + 12$ (viii) $u^2 + 4u + 3$
(ix) $10 + 7w + w^2$
5. (i) $16a + 24$ (ii) $15bc + 35b$ (iii) $3gh + 18g + 2h + 12$
(iv) $5mt + 3m + 20t + 12$ (v) $2n^2 + 17n + 35$ (vi) $6u^2 + 11u + 4$
(vii) $12x^2 + 23x + 5$ (viii) $9 + 18y + 8y^2$ (ix) $7 + 37p + 10p^2$
6. (a) $ax + ay + bx + by + cx + cy$
(b) (i) $f^2 + fg + 5f + 2g + 6$ (ii) $35 + 5k + 12n + kn + n^2$

Exercise 53 (page 126)

1. (i) $gp + 5p, gk - 2g + 5k - 10$ (ii) $hp - 3p, hn + 4h - 3n - 12$
(iii) $7q - wq, 14 - 9w + w^2$
2. (i) $xy + x - 2y - 2$ (ii) $xy - x + 2y - 2$ (iii) $xy - x - 2y + 2$
(iv) $uw - 3u - 7w + 21$ (v) $h^2 - 7h + 10$ (vi) $12 - 7k + k^2$
(vii) $18 - 3m - m^2$ (viii) $f + 4 - df - 4d$ (ix) $7c - 10 - c^2$
3. (i) $2x^2 + 7x - 4$ (ii) $4y^2 - 9y - 9$ (iii) $6up - 3p + 2u - 1$
(iv) $10ab - 15a - 2b + 3$ (v) $2 - 11c + 15c^2$ (vi) $4 - d - 3d^2$
(vii) $2f^2 - 3f - 5$ (viii) $6g^2 - 7g - 3$ (ix) $9 + 3n - 2n^2$
4. (i) $ab + 4a + 3b + 12$ (ii) $c^2 - 12c + 35$ (iii) $12 - 2d + 6f - df$
(iv) $km - 3k + m - 3$
5. (i) $x^2 + 7x + 10, 5$ (ii) 4 (iii) 3 (iv) 3.6
6. (i) 5 (ii) 4 (iii) 2 (iv) 3
7. (i) $x^2 + 6x + 9$ (ii) $a^2 + 2ab + b^2$ (iii) $y^2 + 10y + 25$,
$p^2 + 14p + 49$ (iv) $64 + 16t + t^2, 1 + 12w + 36w^2$
8. (i) $x^2 - 10x + 25$ (ii) $y^2 - 8y + 16, w^2 - 14w + 49$
(iii) $9 - 6p + p^2, a^2 - 2ab + b^2$
9. (i) $a^2 - 25$ (ii) $b^2 - 9$ (iii) $4 - c^2$ (iv) $49 - d^2$ (v) $x^2 - y^2$
(vi) $9f^2 - 1; g^2 - 64, h^2 - 81$
10. (i) $n - 5$ (ii) $p + 8, p - 8$ (iii) $t + 4, t - 4$ (iv) $6 + y, 6 - y$
11. (i) $ab + a + b + 1$; 1581 (ii) $kn - k - 2n + 2$; 2842 (iii) 1386
(iv) 551 (v) 4779 (vi) 1.08 (vii) 2744 (viii) 23.36
12. 1225, 5625, 7225, 20.25, 6.25

Exercise 54 (page 130)

1. (i) $a^2 + 6a + 9$ (ii) $b^2 + 10b + 25$ (iii) $c^2 + 16c + 64$
(iv) $d^2 - 8d + 16$ (v) $e^2 - 14e + 49$ (vi) $f^2 - 20f + 100$
(vii) $16 - 8d + d^2$ (viii) $25 + 10b + b^2$ (ix) $100 - 20f + f^2$
2. (i) $g^2 - 25$ (ii) $h^2 - 9$ (iii) $k^2 - 49$ (iv) $16 - m^2$ (v) $36 - n^2$
(vi) $1 - p^2$ (vii) $9r^2 - 1$ (viii) $25t^2 - 1$ (ix) $1 - 16x^2$
3. (i) $9x^2 + 6xy + y^2$ (ii) $c^2 + 14cd + 49d^2$ (iii) $f^2 + 4fg + 4g^2$
(iv) $h^2 - 8hk + 16k^2$ (v) $25m^2 - 10mn + n^2$ (vi) $9p^2 - 6pq + q^2$
(vii) $r^2 - 64t^2$ (viii) $v^2 - 25w^2$ (ix) $4x^2 - y^2$
4. (i) $c^2d^2 - 9$ (ii) $25 - 10fg + f^2g^2$ (iii) $4h^2 + 12hm + 9m^2$
(iv) $9k^2 - 12kn + 4n^2$ (v) $25p^2 + 20pr + 4r^2$ (vi) $t^6 + 2t^3 + 1$
(vii) $x^6 - 1$ (viii) $9v^2 - 16w^2$ (ix) $4 - y^4$
5. (i) 441 (ii) 1521 (iii) 2704 (iv) 2304 (v) 40 401 (vi) 39 601
(vii) 96.04 (viii) 388.09
6. (i) 8091 (ii) 15.96 (iii) 2436 (iv) 63.84 (v) 39 975 (vi) 0.089 6
7. (i) 25.30 (ii) 1.10 (iii) 4.16 (iv) 16.08
8. (i) 24.70 (ii) 8.70 (iii) 15.84 (iv) 48.86
9. (i) 5 (ii) 7 (iii) $c - 9$ (iv) $d - 10$
(v) $f + 5, f - 5$ (vi) $g + 9, g - 9$
10. (i) $p^2 + 12p + 36$ (ii) $v^2 - 22v + 121$ (iii) $x^2 + 24x + 144$
(iv) $16 - 24y + 9y^2$ (v) $w^2 + 10w + 25$
11. (i) $3 + h$ (ii) $1 - 6k$ (iii) $m^3 - 5$ (iv) $1 + n, 1 - n$
(v) $3p + 1, 3p - 1$ (vi) $q + 4, q - 4$ (vii) $2t - 3$ (viii) $3x - 5y$
12. A and D are identities; for B, $x = 4$; for C, $x = 3$
13. (ii) $\frac{1}{2}$

Exercise 55 (page 133)

1. $3(a + b)$
2. $c(d - e)$
3. $2(f + 5)$
4. $3(2 - g)$
5. $m(h + 2)$
6. $n(n + 5)$
7. $p(4 - p)$
8. $r(r + 1)$
9. $t(1 - t)$
10. $5a(b + c)$
11. $2c(d - 3e)$
12. $5f(2f - 1)$
13. $3g(1 + 2g)$
14. $5h(2k - 3m)$
15. $p^4(p^2 + 1)$
16. $t^2(t^3 - 2)$
17. $a(b + c + d)$
18. $2(2f - 3g + 5h)$

Exercise 56 (page 133)

1. $(x + 5)(x - 5)$
2. $(y + 6)(y - 6)$
3. $(c + 9)(c - 9)$
4. $(8 + d)(8 - d)$
5. $(3 + f)(3 - f)$
6. $(1 + g)(1 - g)$
7. $(3h + 2)(3h - 2)$
8. $(7 + 3k)(7 - 3k)$
9. $(6m + 1)(6m - 1)$
10. $(1 + 2n)(1 - 2n)$
11. $(p + \frac{1}{3})(p - \frac{1}{3})$
12. $(q + \frac{2}{5})(q - \frac{2}{5})$
13. $(r + 1\frac{1}{2})(r - 1\frac{1}{2})$
14. $(1 + t^3)(1 - t^3)$
15. $(w^8 + 5)(w^8 - 5)$
16. 25, 5, 5
17. $3(y + 2)(y - 2)$
18. $5(z + 3)(z - 3)$
19. $a(b + 3)(b - 3)$
20. $c(c + d)(c - d)$
21. $f(7 + 5f)(7 - 5f)$

22. (i) 4600 (ii) 1200 (iii) 10 600 (iv) 13 200 (v) 0.4 (vi) 10.8 (vii) 108 (viii) 0.17
23. (i) 8 (ii) 9 (iii) 2 (iv) 2

Exercise 57 (page 135)

1. (i) $x^2 + 7x + 12$ (ii) $y^2 + 7y + 6$ (iii) $p^2 + 15p + 50$ (iv) $q^2 + 11q + 28$ (v) $k^2 + 17k + 66$ (vi) $n^2 + 13n + 36$
2. (i) $a^2 - 8a + 15$ (ii) $b^2 - 3b + 2$ (iii) $c^2 - 7c + 12$ (iv) $d^2 - 8d + 12$ (v) $e^2 - 10e + 9$ (vi) $f^2 - 13f + 40$
3. (i) $g^2 + 2g - 8$ (ii) $h^2 + 3h - 28$ (iii) $k^2 - 3k - 28$ (iv) $m^2 - 6m - 27$ (v) $n^2 + 3n - 40$ (vi) $p^2 - 3p - 40$ (vii) $q^2 - 16$ (viii) $t^2 - 1$ (ix) $u^2 - 3u - 18$
4. (a) (i) plus signs (ii) minus and plus signs (b) (i) 5, 1 (ii) 3, 1 (iii) $n + 7, n + 1$ (iv) $p - 3, p - 2$
5. (a) minus signs (b) (i) 3, 1 (ii) 1, 5 (iii) $h + 7, h - 1$ (iv) $m - 3, m + 2$

Exercise 58 (page 136)

1. (i) $(x + 3)(x + 1)$ (ii) $(y + 2)(y + 1)$ (iii) $(p + 11)(p + 1)$ (iv) $(q - 2)(q - 1)$ (v) $(r - 5)(r - 1)$ (vi) $(t - 13)(t - 1)$
2. (ii) and (iii) cannot be factorised
(i) $(x + 7)(x + 1)$ (iv) $(x - 7)(x - 1)$
3. (a) $x^2 + 7x + 10$, $x^2 + 11x + 10$. First
(b) $y^2 + 13y + 12$, $y^2 + 8y + 12$, $y^2 + 7y + 12$. Second
(c) $u^2 + 25u + 24$, $u^2 + 14u + 24$, $u^2 + 11u + 24$, $u^2 + 10u + 24$. Last
4. (a) 1, 18; 2, 9; 3, 6; 6, 3; 9, 2; 18, 1; $(x + 2)(x + 9)$, $(x + 3)(x + 6)$
(b) 1, 30; 2, 15; 3, 10; 5, 6; 6, 5; 10, 3; 15, 2; 30, 1; $(p - 5)(p - 6)$, $(q - 3)(q - 10)$
(c) 1, 16; 2, 8; 4, 4; 8, 2; 16, 1; $(n + 4)(n + 4)$, $(k + 2)(k + 8)$
(d) 1, 36; 2, 18; 3, 12; 4, 9; 6, 6; 9, 4; 12, 3; 18, 2; 36, 1; $(m - 4)(m - 9)$, $(t - 1)(t - 36)$
5. (i) $(a + 3)(a + 4)$ (ii) $(b + 2)(b + 6)$ (iii) impossible (iv) $(d - 4)(d - 2)$ (v) impossible (vi) $(f - 1)(f - 8)$ (vii) $(g + 2)(g + 3)$ (viii) impossible (ix) $(k + 3)(k + 5)$
6. (b) 3; $(p + 1)(p + 2)$ (c) 16, $(k - 1)(k - 15)$; 8, $(k - 3)(k - 5)$

Exercise 59 (page 137)

1. (i) $(a + 1)(a - 3)$ (ii) $(b + 3)(b - 1)$ (iii) $(c + 7)(c - 1)$ (iv) $(d + 1)(d - 7)$ (v) $(e + 1)(e - 11)$ (vi) $(f + 1)(f - 5)$ (vii) $(g + 2)(g - 1)$ (viii) $(h + 1)(h - 4)$ (ix) $(k + 6)(k - 1)$

2. Four pairs (i) $(x+2)(x-5)$ (ii) $(y+5)(y-2)$ (iii) $(u+10)(u-1)$ (iv) $(w+1)(w-10)$
3. (i) $(k+3)(k-4)$ (ii) $(m+4)(m-3)$ (iii) $(n+2)(n-6)$ (iv) $(p+6)(p-2)$; $x^2+11x-12$, $x^2-11x-12$
4. (i) $(a+2)(a-4)$ (ii) $(b+1)(b-8)$ (iii) impossible (iv) $(d+4)(d-2)$
5. (i) $(f+3)(f-6)$ (ii) impossible (iii) $(h+2)(h-9)$ (iv) $(k+1)(k-18)$ (v) impossible (vi) $(n+6)(n-3)$
6. (i) $(p+1)(p-9)$ (ii) $(r+3)(r-3)$ (iii) $(u-3)(u-3)$ (iv) $(w-1)(w-9)$ (v) impossible (vi) $(y+3)(y+3)$

Exercise 60 (page 138)

1. (i) $12+4a-a^2$ (ii) $21-4b-b^2$ (iii) $50-5c-c^2$ (iv) $6-5d-d^2$ (v) $36-9e-e^2$ (vi) $18+7f-f^2$
2. (i) 7, 2 (ii) 10, 1 (iii) $(3+k)(5-k)$ (iv) $(8+m)(1-m)$
3. (i) $a^2+2ab-24b^2$ (ii) $c^2-10cd+21d^2$ (iii) e^6-4e^3-5 (iv) $14+9f^2+f^4$ (v) $12g^2-g-1$ (vi) $21h^2-10hn+n^2$
4. (i) 5, 1 (ii) $5u$, $2u$ (iii) $5y$, y (iv) $5g$, $2g$
5. (i) $(3+x)(2+x)$ (ii) $(2-y)(5-y)$ (iii) $(4+u)(2-u)$ (iv) $(3+w)(4-w)$ (v) $(1-p)(8-p)$ (vi) $(2+r)(6-r)$
6. (i) $(1-a)(1-4a)$ (ii) $(1+3b)(1-b)$ (iii) $(c+2d)(c-5d)$ (iv) $(3f-g)(f-g)$ (v) $(h^2+5)(h^2+1)$ (vi) $(1+2k^3)(1-3k^3)$
7. (i) $n(n-5)$ (ii) $k(k+8)$ (iii) $p(6p-1)$ (iv) $q(10+q)$ (v) $x^2(x+4)$ (vi) $3y(y-2)$
8. (i) $(16h-1)(h-1)$ (ii) impossible (iii) $(4m+1)(4m-1)$ (iv) $(4p+1)^2$ (v) impossible (vi) $(8y+1)(2y-1)$
9. (a) (i) $+3a-10$, $+5$, -2 (ii) 9, 9; 9, 3, 3
   (b) (i) $c(c+3)(c+2)$ (ii) $d^2(d+1)(d-7)$
   (iii) $(f^2+4)(f+2)(f-2)$ (iv) $5(g+1)(g-3)$
   (v) $3h(1-h)(1-4h)$ (vi) $(1+n^2)(1+n)(1-n)$

Exercise 61 (page 139)

1. (a), (b), (d), (e)
3. (a), (b), (d), (e)
5. 3, 2; outside
6. (a) 2 (b) 3 (c) 3 (d) 7 (e) 9 (f) 1
8. (a) and (b); (c) and (e)
9. 4

Exercise 62 (page 143)

1. (a) 2 odd, 2 even (b) 2 odd (c) 4 odd, 1even (d) 3 even
2. Arcs: (a) 4 (b) 3 (c) 7 (d) 6. Regions: (a) 2 (b) 3 (c) 4 (d) 5
4. $N+R-A=2$

Exercise 63 (page 145)

1. (a), (b), (c) and (g) are traversable

Exercise 64 (page 147)

1. 0.64
2. 0.17, 0.34, 0.50, 0.64, 0.77, 0.87, 0.94, 0.98; 0.57, 0.82
3. (ii) 0.26, 0.39, 0.62, 0.83, 0.97 (iii) 26°, 34°, 46°, 64°, 72°
4. (i) 13 cm (ii) 5.2 cm (iii) 7.5 cm
5. (i) 23 cm (ii) 6.4 cm (iii) 3.6 cm
6. (i) 37° (ii) 49° (iii) 27°

Exercise 65 (page 149)

1. 0.77
2. 0.98, 0.94, 0.87, 0.77, 0.64, 0.50, 0.34, 0.17; 0.82, 0.57
3. (iii) 0.91, 0.85, 0.71, 0.45, 0.22 (iv) 14°, 38°, 55°, 68°, 83°
4. (i) 18 cm (ii) 2.3 m
5. (i) 66° (ii) 38°

Exercise 66 (page 150)

1. (i) 0.5621 (ii) 0.9728 (iii) 0.0993 (iv) 0.3605 (v) 0.5842 (vi) 0.9389
2. (i) 0.9379 (ii) 0.5000 (iii) 0.1115 (iv) 0.5060 (v) 0.5055 (vi) 0.9186
3. (i) 17° (ii) 34°36′ (iii) 34°37′ (iv) 58°42′ (v) 58°44′ (vi) 48°28′
4. (i) 73° (ii) 55°24′ (iii) 55°29′ (iv) 28°54′ (v) 28°57′ (vi) 68°16′
5. 4.69 cm, 8.83 cm **6.** 16.4 cm, 11.5 cm
7. 4.81 cm, 6.39 cm **8.** 4.30 cm, 11.2 cm
9. 3.81 cm, 2.40 cm **10.** 2.69 cm, 0.791 cm
11. 41.3° **12.** 48.6° **13.** 31.8° **14.** 63.6° **15.** 30°
16. 40.5° **17.** 5.47 cm **18.** 118° **19.** 53.1°, 17 cm
20. 9.44 cm **21.** 10.3 cm
22. (i) 0.6428, 0.9848; No (ii) Neither
23. 0.6157, 0.6157 (i) 68° (ii) 56.3°

Exercise 67 (page 154)

1. 72.5° **2.** 7.52 m **3.** 79.5 m **4.** 6.3° **5.** 50.1°
6. (i) 8.45 km (ii) 18.1 km **7.** (i) 23.5 km S (ii) 32.4 km W
8. 68.5° **9.** (i) 72° (ii) 11.8 cm (iii) 8.1 cm (iv) 239 cm^2
10. (i) 9.83 cm (ii) 6.88 cm (iii) 33.8 cm^2

Revision Paper B1 (page 155)

1. (a) 0.34, 6.8, 0.068 (b) $\frac{9}{20}, \frac{21}{25}, \frac{7}{8}, \frac{12}{125}$ (c) 0.28, 0.65, 0.94, 0.095
2. (a) 2 (b) 1, -3 **3.** 52, 58
4. (b) (i) $ab + 5a + 3b + 15$ (ii) $c^2 + 8c + 12$
(iii) $d^2 - 5d - 24$ (iv) $35 - 12f + f^2$
5. (a) I and III (b) 12 cm
6. (i) 11 h, 8 h (ii) $A = 34 - 2S$ (iii) 6 years, 14 years
7. (i) 4.69 cm (ii) 52°8′ **8.** 30 to 40

Revision Paper B2 (page 156)

1. (a) 900, 0.09, 0.0009 (b) 20, not possible, 0.2, 0.5, not possible
(c) 58.3, 18.4, 0.583, 0.184
2. (i) $5(a - b)$ (ii) $c(c - 1)$ (iii) $(d + 4)(d - 4)$
(iv) $(1 + f)(1 - f)$ (v) $(g + 3)(g + 1)$ (vi) $(h - 5)(h - 1)$
3. (i) 4.9×10^{-3} (ii) 9×10^{-4} (iii) 4.4×10^{-3} (iv) 6×10^{-3}
(v) 9×10^2

4. 27, 27. **5.** 220 cm, $454\frac{6}{11}$ **6.** £$\frac{x}{6}$, £$\frac{x}{9}$, 9

7. 128 m, 21.8° (21°48′) **8.** $y = x$; (7, 5); (59, 17)

Revision Paper B3 (page 157)

1. (a) (i) (b) (i) and (iii) **2.** (b) $\frac{2}{5}$
3. (a) (i) $a^2 - 49$ (ii) $2b^2 - 7b + 3$ (iii) $15 + 7c - 2c^2$
(iv) $d^2 - 8d + 16$ (b) 5
4. (i) 238 cm^2 (ii) 3.73 cm
5. (i) Impossible, $\begin{pmatrix} -6 & 27 \\ 5 & 16 \end{pmatrix}$, $\begin{pmatrix} -8 & 24 \\ 2 & 7 \\ -10 & 21 \end{pmatrix}$, $\begin{pmatrix} 6 & -6 & 5 \\ 10 & 6 & -9 \end{pmatrix}$, $\begin{pmatrix} 8 & 4 \\ 8 & 4 \end{pmatrix}$

(ii) $\begin{pmatrix} 2 & 1 \\ 2 & 1 \end{pmatrix}$ (iii) $\begin{pmatrix} -8 & 28 \\ 7 & 11 \end{pmatrix}$

6. 18 fivepenny coins, 15 twopenny

7. (i) 4.48 cm (ii) 50°29′ **8.** 14.1; $y = 2x + \dfrac{400}{x}$

Revision Paper B4 (page 159)

1. (i) Impossible (ii) $(b + 3)(b - 3)$ (iii) $(c + 9)(c - 1)$
(iv) impossible (v) $(e - 9)(e - 1)$ (vi) impossible

2. (i) $n = \frac{1}{2}r + 2$ (ii) $P = \dfrac{100I}{R}$ (iii) $x = \sqrt{100 - y}$

(iv) $r = \sqrt{\dfrac{A}{\pi}}$ (v) $h = \dfrac{T^2}{4}$

3. (i) $\tan\theta = \dfrac{y}{x}$ and $\sin\theta = \dfrac{y}{h}$ (ii) 58°0′ (iii) 36°52′, 8, 24

4. (a) $-64, \dfrac{27}{125}, 0.008, a^{12}$ (b) $-3, \frac{1}{2}, 0.1, b^2$

(c) 21.7, 1.41

5. (a) (ii) and (iv) impossible; (i) 5 (iii) 12 (v) 20 (b) 75°

6. Length = 3 × breadth; yes; $y = \frac{1}{3}x$

7. 72 m², 41 m² **8.** 2, 2, 2.3 approx.

Revision Paper B5 (page 161)

1. (a) $a^2 - b^2$ (i) 1596 (ii) 99.91

(b) $a^2 + 2ab + b^2$; 16.32, 4.12

2. (a) (i) $\dfrac{a}{2c}$ (ii) d^2 (iii) $\frac{7}{9}$ (iv) $\dfrac{h}{2}$

(b) (i) $\dfrac{x}{12}$ (ii) $\dfrac{3y + 2}{y^2}$ (iii) $\dfrac{2}{3}$ (iv) $\dfrac{ad}{bc}$

3. £20, £18 **4.** 2.65 m, 3.41 m

5. (a) (i) 15p (ii) 60p (iii) 440 g (iv) 102 minutes

(b) (i) 15% (ii) 34%

7. Points total for each team, (13, 11, 10, 12). B is best.

8. (i) 13 cm (ii) 20 cm (iii) 252 cm²

Exercise 68 (page 164)

1. 1.7542, $\bar{1}$.7542, $\bar{2}$.7542, $\bar{4}$.7542

2. 3.9294, 2.9294, 1.9294, 0.9294, $\bar{1}$.9294, $\bar{2}$.9294

3. (i) $\bar{1}$.4116, $\bar{2}$.4116, $\bar{4}$.4116 (ii) $\bar{1}$.9253, $\bar{4}$.9253, 3.9253

(iii) $\bar{2}$.7016, $\bar{1}$.7016, 1.7016

Exercise 69 (page 165)

1. (i) $\bar{3}$ (ii) $\bar{2}$ (iii) $\bar{1}$ (iv) $\bar{4}$ (v) 1 (vi) $\bar{3}$ (vii) $\bar{1}$ (viii) 3

2. $\bar{1}$.8603, $\bar{3}$.8603, 1.8603, $\bar{2}$.8603 **3.** $\bar{2}$.8344, 0.8344, $\bar{4}$.8344, 3.8344

4. 1.7042, $\bar{1}$.7042, $\bar{2}$.7042, $\bar{4}$.7042 **5.** 2.4771, $\bar{2}$.4771, $\bar{4}$.4771, $\bar{1}$.4771

6. 1.841, 0.1841, 0.018 41, 0.001 841

7. 68.39, 0.6839, 0.006 839, 0.000 068 39

8. 0.052 29, 5.229, 522.9, 0.5229

9. 0.1239, 0.001 239, 12 390, 0.000 123 9

Exercise 70 (page 166)

1. (i) 3 (ii) $\bar{5}$ (iii) $\bar{3}$ (iv) 5 (v) 0 (vi) $\bar{7}$
(vii) 7 (viii) $\bar{2}$ (ix) 6 (x) $\bar{6}$
2. (i) 1.7 (ii) $\bar{3}.7$ (iii) $\bar{1}.3$ (iv) $\bar{3}.3$ (v) $\bar{2}.4$
(vi) 2.6 (vii) 0.3 (viii) $\bar{3}.3$ (ix) $\bar{8}.7$ (x) $\bar{4}.4$
3. 78.2 **4.** 8.62 **5.** 0.188 **6.** 0.215/6
7. 48.2 **8.** 0.0795 **9.** 0.008 51 **10.** 0.003 53
11. 58.7 **12.** 0.501 **13.** 8.20 **14.** 1.21

Exercise 71 (page 167)

1. (i) 3, $\bar{7}$, $\bar{3}$, $\bar{6}$, $\bar{8}$, (ii) 4, 3, 9, 2, 6
(iii) $\bar{5}$, $\bar{1}$, 1, 5, $\bar{5}$ (iv) $\bar{1}$, 1, $\bar{3}$, 3, $\bar{1}$
2. (i) $\bar{4}.2$ (ii) $\bar{3}.4$ (iii) 2.2 (iv) 3.7 (v) 5.5
(vi) $\bar{3}.6$ (vii) $\bar{2}.7$ (viii) 1.8 (ix) $\bar{7}.8$ (x) 5.7
3. (i) $\bar{2}.4$ (ii) 1.7 (iii) $\bar{2}.8$ (iv) 1.7 (v) $\bar{6}.2$
(vi) 7.7 (vii) $\bar{9}.7$ (viii) 7.6 (ix) 1.4 (x) $\overline{11}.8$
4. 0.130 **5.** 0.570 **6.** 0.337 **7.** 0.006 95
8. 84.6 **9.** 88.6 **10.** 7.61 **11.** 61.6
12. $1.77/8 \times 10^{-6}$ **13.** 4.09×10^{-7} **14.** 1.32×10^{-7} **15.** 9.77×10^{-14}

Exercise 72 (page 168)

1. (i) $\bar{2}.6$ (ii) $\bar{5}.2$ (iii) $\overline{10}.1$ (iv) $\bar{1}.6$
(v) $\overline{17}.2$ (vi) $\bar{1}.0$ (vii) $\bar{3}.72$ (viii) $\overline{14}.68$
2. 0.014 9 **3.** 0.802 **4.** 0.188 **5.** 0.418 **6.** 16 900
7. 0.016 9 **8.** 4920 **9.** 0.492 **10.** 0.047 9 **11.** 0.415
12. $1.77/8 \times 10^{-6}$ **13.** 4.09×10^{-7} **14.** 1.32×10^{-7} **15** 9.77×10^{-14}

Exercise 73 (page 169)

1. (i) 716 (ii) 71.6 (iii) 5290 (iv) 1480
2. 0.895 **3.** 0.415 **4.** 0.853 **5.** 0.480 **6.** 0.167
7. 0.975 **8.** 0.431 **9.** 0.333 **10.** 0.063 0 **11.** 0.779
12. 0.518

Exercise 74 (page 169)

1. 38.4 **2.** 0.0142 **3.** 0.272 **4.** 0.865 **5.** 0.0331
6. 32.7 **7.** 0.968 **8.** 0.590/1 **9.** 0.625 **10.** 0.0899
11. 0.701 m^2 **12.** 0.921 m **13.** 0.0748 mm **14.** £351 **15.** 0.006 053

Exercise 75 (page 171)

2. (a) $12 \times 7 = 84, 7 \times 12 = 84$ (b) $2^4 = 16$
(c) $(a + b)(a - b) = a^2 - b^2$ (d) $\{p, q\} \cap \{q, r\} = \{q\}$
4. (a) 108, 300 (b) 49, 81 (c) 3, 1.8
7. 314 **8.** (a) 45° (b) 13

Exercise 76 (page 175)

1. (i), (ii) and (v) are suitable. (iii) Is x greater than y? (iv) Should I turn left at the T junction? (vi) Did Tom win the point? (vii) Did the coin show a head? (viii) Do you like football?
2. £13.20, £39.50 **3.** 42 percs, 66 percs, 187 percs, 55 percs
4. 20 is a factor of 840

Exercise 77 (page 178)

1. Multiples of 7 from 56 to 91
2. It forms the sum of the odd numbers less than 20

Exercise 78 (page 182)

1. Scalars: (a), (c), (d), (f). Vectors: (b), (e)
2. (b), (c), (d), (f)
7. C and G, D and K, E and H

Exercise 79 (page 185)

1. (i) $\begin{pmatrix}1\\5\end{pmatrix}, \begin{pmatrix}2\\2\end{pmatrix}, \begin{pmatrix}3\\-4\end{pmatrix}, \begin{pmatrix}-1\\6\end{pmatrix}, \begin{pmatrix}0\\2\end{pmatrix}$ (ii) $\vec{PR}, \vec{QR}, \vec{ST}, \vec{PT}, \vec{TQ}$
2. (ii) $\begin{pmatrix}4\\3\end{pmatrix}, \begin{pmatrix}2\\0\end{pmatrix}, \begin{pmatrix}1\\-3\end{pmatrix}, \begin{pmatrix}6\\-4\end{pmatrix}, \begin{pmatrix}-5\\1\end{pmatrix}$ (iii) $\vec{TR}, \vec{ZX}, \vec{YR}, \vec{XT}$
3. $\vec{AB} = \begin{pmatrix}4\\1\end{pmatrix}, \vec{AC} = \begin{pmatrix}1\\4\end{pmatrix}$. Same. Equal in length, same angle to N and E
4. $\begin{pmatrix}6\\1\end{pmatrix}, \begin{pmatrix}7\\5\end{pmatrix}, \begin{pmatrix}2\\3\end{pmatrix}, \begin{pmatrix}1\\6\end{pmatrix}$
5. (iii) Each is $\begin{pmatrix}2\\3\end{pmatrix}$. They are equivalent. OB = AC, OB parallel to AC.
(iv) Parallelogram
6. (i) $\begin{pmatrix}0\\3\end{pmatrix}, \begin{pmatrix}3\\2\end{pmatrix}, \begin{pmatrix}2\\0\end{pmatrix}, \begin{pmatrix}1\\4\end{pmatrix}, \begin{pmatrix}4\\3\end{pmatrix}$ (ii) $\vec{PQ}, \vec{TR}$
(iii) $\begin{pmatrix}1\\2\end{pmatrix} \begin{pmatrix}-4\\-1\end{pmatrix} \begin{pmatrix}-1\\-4\end{pmatrix} \begin{pmatrix}2\\6\end{pmatrix}$
7. (2, 0), (5, −1), (4, −3), (0, −6)

Exercise 80 (page 187)

2. (i) $\begin{pmatrix}6\\6\end{pmatrix}$ (ii) $\begin{pmatrix}1\\8\end{pmatrix}$ (iii) $\begin{pmatrix}4\\3\end{pmatrix}$ (iv) $\begin{pmatrix}4\\6\end{pmatrix}$

4. (i) $\begin{pmatrix}5\\9\end{pmatrix}$ (ii) $\begin{pmatrix}1\\6\end{pmatrix}$ (iii) $\begin{pmatrix}0\\5\end{pmatrix}$ (iv) $\begin{pmatrix}5\\9\end{pmatrix}$ (v) $\begin{pmatrix}7\\0\end{pmatrix}$ (vi) $\begin{pmatrix}2\\-4\end{pmatrix}$

5. (i) 1, 3 (ii) 8, 2 (iii) 3, -3 (iv) 1, -1 (v) 5, 1 (vi) 4, 10

6. $\begin{pmatrix}16\\8\end{pmatrix}, \begin{pmatrix}28\\14\end{pmatrix}, \begin{pmatrix}2\\1\end{pmatrix}$

7. $\begin{pmatrix}6\\8\end{pmatrix}, \begin{pmatrix}18\\9\end{pmatrix}, \begin{pmatrix}48\\-32\end{pmatrix}, \begin{pmatrix}2\\1\end{pmatrix}, \begin{pmatrix}6\\-4\end{pmatrix}, \begin{pmatrix}18\\0\end{pmatrix}, \begin{pmatrix}27\\21\end{pmatrix}$

8. (i) $\begin{pmatrix}-2\\-5\end{pmatrix}, \begin{pmatrix}3\\-1\end{pmatrix}, \begin{pmatrix}-4\\6\end{pmatrix}$ (ii) $\begin{pmatrix}-2\\-5\end{pmatrix}, \begin{pmatrix}3\\-1\end{pmatrix} \begin{pmatrix}-4\\6\end{pmatrix}$

9. $\begin{pmatrix}-3\\-4\end{pmatrix}, \begin{pmatrix}-6\\-3\end{pmatrix}, \begin{pmatrix}-12\\8\end{pmatrix}$

10. (ii) $\begin{pmatrix}-5\\-1\end{pmatrix}, \begin{pmatrix}-3\\-2\end{pmatrix}, \begin{pmatrix}1\\-4\end{pmatrix}, \begin{pmatrix}-6\\-6\end{pmatrix}, \begin{pmatrix}0\\2\end{pmatrix}, \begin{pmatrix}-4\\0\end{pmatrix}$

Exercise 81 (page 190)

1. $\begin{pmatrix}1\\4\end{pmatrix}, \begin{pmatrix}5\\0\end{pmatrix}, \begin{pmatrix}6\\4\end{pmatrix}, \begin{pmatrix}3\\2\end{pmatrix}, \begin{pmatrix}6\\4\end{pmatrix}$ **4.** $\begin{pmatrix}7\\3\end{pmatrix}, \begin{pmatrix}2\\5\end{pmatrix}, \begin{pmatrix}-2\\5\end{pmatrix}, \begin{pmatrix}2\\5\end{pmatrix}$

7. (3, 0), (2, 2); $\begin{pmatrix}-1\\2\end{pmatrix}, \begin{pmatrix}-2\\4\end{pmatrix}$; $\overrightarrow{P'Q'} = 2\overrightarrow{PQ}$;
ratio 1:2; directions the same.

9. (3, 0), (5, 3), (2, 5), (0, 2); $\begin{pmatrix}2\\3\end{pmatrix}, \begin{pmatrix}2\\3\end{pmatrix}$; Same; Parallelogram

12. (i) -3, 7 (ii) 1, -1 (iii) 11, 8 (iv) 3,2

Exercise 82 (page 193)

1. 6, 0, 7, -6, 0, 9, 0, 0
2. 2, 4, 1, 0, 0, 0, $\frac{1}{3}$, $\frac{4}{3}$, 3.6, 0, 0, 4
3. (i) 3 (ii) 0 (iii) $\frac{1}{9}$ (iv) any number (v) 3
(vi) any number (vii) $\frac{3}{4}$ (viii) 0 (ix) 0
4. (i) $a = 0$ (ii) $b = 0$ (iii) $b =$ any number
5. (i) 18 (ii) 0 (iii) -2 (iv) 0 (v) 10
6. 10, 4, 0, -2, -2, 0, 4; 3, 6
7. 1.25, -0.96 **8.** 0, -2, -2, 0
9. 3, 0, -1, 0, 3, 8; -2, -4
10. 8, 0, -6, -10, -6, 0; 3, -4
11. 5, 7 **12.** -2, -6 **13.** 4, -3 **14.** 0, 8
15. 6, 0, -4, -6, -6, -4, 0, 6; -2, 3 **16.** 2, -1

Exercise 83 (page 195)

1. 2, 5 **2.** 1, 3 **3.** $-3, -4$ **4.** $-1, -6$ **5.** $5, -2$
6. $1, -3$ **7.** 0, 9 **8.** $3, -3$ **9.** $5, -7$ **10.** $5, -3$
11. $5, -4$ **12.** $5, -6$ **13.** $2, -5$ **14.** $6, -2$ **15.** $8, -1$
16. $2, -7$ **17.** 1, 2 **18.** $2, -3$
19. $7, 0, -5, -8, -9, -8, -5, 0, 7; 4, -2$
20. $8, 3, 0, -1, 0, 3, 8; 4, 6$
21. No **22.** $-1, -5$ **23.** 3, 9 **24.** $3, -2$ **25.** 3
26. $4, -5$ **27.** $-1, -4$

Exercise 84 (page 196)

1. $n^2 + (n + 3)^2 = 65; n^2 + 3n - 28 = 0$; 4 and 7 (or -4 and -7)
2. 4, 12 (or -4 and -12) **3.** $(x + 6)$ m; $x(x + 6)$ m^2; 16 m, 10 m
4. $(n + 4)$y; $n(n + 4) = 96$; 12 years, 8 years
5. $(y + 7)$ cm, $(y + 9)$ cm; $y^2 - 4y - 32 = 0$, 8, -4; 8, 15, 17 cm
6. $x^2 - 16x + 48 = 0$; 4, 12; 12 cm **7.** 1 s, 5 s; 2s, 4s
8. $3, -4; (3, 5), (-4, -2)$ **9.** $(-1, -4), (5, 20)$
10. $(2, 5), (-5, -2)$

Exercise 85 (page 199)

1. At 28 s, 42 m; At 40 s. **3.** (i) 2 m/s (ii) 20 s (iii) $1\frac{1}{2}$ m/s
4. (i) 60 m, 15 m/s (ii) 20 m, 4 m/s (iii) 10 m/s, $1\frac{2}{3}$ m/s
(iv) 54 km/h, 6 km/h (v) 30, 60
7. (i) 30 min (ii) 24 km/h (iii) 20 min (iv) 40 min (v) 18 km/h
(vi) 9 km (vii) 10.10 a.m. (viii) 6 km/h (ix) 7 km
(x) 10.20 a.m. and 11.07 a.m.
8. (i) 12.00 (ii) 45 km (iii) 15 min (iv) 30 min (v) 10 km
(vi) 20 km/h (vii) upstream (viii) $2\frac{3}{4}$ h (ix) 20 km/h (x) 10.45
9. 14.00; 13.33, 33 km from home
10. 15.15 at 90 km from Junction 5 and 17.06 at 238 km from Junction 5
12. 10.42, 10.1 km from A
13. (i) 20 m, 120 m, 420 m (ii) 20 m/min, 60 m/min, 148 m/min
(iii) 108 m/min (iv) 3 min

Exercise 86 (page 206)

1. (i) 0.7 m/s^2 (ii) 0.9 m/s^2 (iii) 4km/h per s
2. At 6 min and $42\frac{1}{2}$ min; $6\frac{2}{3}$, 2, $4\frac{1}{6}$ km/h per min
3. 0.4 m/s^2, 0.5 m/s^2; 224 m
4. (i) At 22 s and 66 s (ii) 2.5, 1.25, $1\frac{2}{3}$ m/s^2 (iii) 3125 m
6. (i) 13 s (ii) 500 m

Exercise 87 (page 208)

1. 7 m, 2 m, 36 m, 64 m **2.** 5 cm, 16 cm, 5 cm, 32 cm
3. 3600 g, 5900 g, 9300 g, 300 g
4. 5.84, 3.73, 0.60, 1.01 **5.** 35.8, 23.3, 4.8, 12.0
6. 5.27, 7240, 9030, 0.0727 **7.** 15, 6700, 850, 0.0053
8. (i) 33.5 km and 34.5 km (ii) 25.5 cm and 26.5 cm
(iii) 525 g and 535 g (iv) £3150 and £3250
(v) 5.55 s and 5.65 s (vi) $2\frac{3}{4}$ h and $3\frac{1}{4}$ h
9. 6.335 and 6.345, 0.855 and 0.865, 0.065 and 0.075, 9.025 and 9.035
10. 71.5 and 72.5, 6250 and 6350, 0.525 and 0.535, 4.55 and 4.65
11. (i) 3.14 (ii) 3.142 (iii) 3.1 (iv) 3.1
12. (i) 39.5 cm to 40.5 cm (ii) 35 cm to 45 cm
(iii) 39.75 cm to 40.25 cm
13. 51.5 m to 52.5 m, 51.95 m and 52.05 m

Exercise 88 (page 209)

1. (i) 24 cm^2 (ii) 29.25 cm^2, 19.25 cm^2
2. 80 cm^2, 89.25 cm^2, 71.25 cm^2
3. 432 cm^2, 453.25 cm^2, 411.25 cm^2
4. 81 cm^2, 90.25 cm^2, 72.25 cm^2
5. 42 cm^2, 46.875 cm^2, 37.375 cm^2
6. 113 cm^2, 133 cm^2, 95 cm^2
7. 38 cm, 34 cm; 86 cm, 82 cm
8. (i) 30 cm^2, 20 cm^2 (ii) 29 cm^2, 27 cm^2
9. 18 km, 12 km **10.** 116 km, 101 km
11. 15 m/s; 2.1 km; 16.5 m/s, 13.5 m/s; 2.5 km, 1.8 km
12. 240 cm^3, 307 cm^3, 183 cm^3

Exercise 89 (page 211)

1. $2\frac{1}{5}$, $3\frac{2}{3}$, $2\frac{3}{5}$, $4\frac{1}{3}$; Least distance and greatest time
2. 5.4 km/h, 3.8 km/h **3.** $18\frac{3}{4}$ km/h, $16\frac{1}{4}$ km/h
4. 2 min; Yes by 3 min **5.** (i) 2.43, 1.67 (ii) 2.04, 1.96
6. (i) 31 m/s (ii) 110 km/h; 36 m/s, 130 km/h; 27 m/s, 96 km/h
7. 9.0 g/cm^3, 8.8 g/cm^3

Exercise 90 (page 214)

1 (i) 61 (ii) 3.6 (iii) 1300 (iv) 2.5 (v) 3.7 (vi) 0.27
2. (i) 0.824 (ii) 0.615 (iii) 0.898 (v) 0.371 (v) 0.213 (vi) 0.0828
4. (i) 237 (ii) 7.17 (iii) 250 (iv) 0.600 (v) 8.19 (vi) 4.80
5. (i) 271 (ii) 48.8 (iii) 0.484 (iv) 11.3
6. £217.16 **7.** £1134.20

8. £88.92
9. £7520
10. 2187; 59 049
11. 40 320; 479 001 600
12. (i) 1881 (ii) 3792 (iii) 0.077 08 (iv) 188.48 (v) 26.637 (vi) 321.64

Exercise 91 (page 215)

1. 111 111, 222 222, 333 333, 444 444; 15 873 × 35, 15 873 × 42
2. 121, 12 321, 1 234 321, ...; yes
3. 121, 1331, 14 641, 161 051, 1 771 561; 11^2, 11^3, 11^4
4. 54, 6534, 665 334, 66 653 334, ...; no; 666 665 333 334
5. 333 333 333, 444 444 444; ×45, ×63
6. 36, 64; perfect squares; 35; 1600; 1681
7. 1367, 1369; 1369, 1521; $38^2 - 38 - 37 = 37^2$; $50^2 - 50 - 49 = 49^2$; 2401
8. $0.\dot{0}\dot{9}$, $0.\dot{2}\dot{7}$, $0.\dot{3}\dot{6}$. $0.\dot{5}\dot{4}$, $0.\dot{6}\dot{3}$, $0.\dot{7}\dot{2}$, $0.\dot{8}\dot{1}$, $0.\dot{9}\dot{0}$; They add up to 9
9. $0.\dot{3}$, $0.\dot{6}$, $0.\dot{1}$, $0.\dot{2}$, $0.\dot{4}$, $0.\dot{5}$, $0.\dot{7}$, $0.\dot{8}$
10. $0.\dot{8}5714\dot{2}$; $0.\dot{2}8571\dot{4}$, $0.\dot{4}2857\dot{1}$, $0.\dot{5}7142\dot{8}$, $0.\dot{7}1428\dot{5}$; $\frac{2}{7}$ and $\frac{5}{7}$, $\frac{3}{7}$ and $\frac{4}{7}$
11. $0.\dot{0}7692\dot{3}$, $0.\dot{1}5384\dot{6}$, $0.\dot{2}3076\dot{9}$, ...; $\frac{1}{13}$ and $\frac{12}{13}$, $\frac{2}{13}$ and $\frac{11}{13}$
12. 34, 55, 89, 144, 233, 377, 610, 987, 1597, 2584, 4181, 6765, 10 946; Golden number is 1.618 033 989 ....

Exercise 92 (page 217)

1. 0.9, 0.35, 0.6, 0.2857 ..., 0.125, 0.1875, 0.7647 ..., 0.5, 0.744 ..., 0.38
2. 860, 836, 3300, 3551
3. 8.5, 4.7, $4.0\dot{5}$, 3
4. 1.26, 1.482, 0.18, 0.4
5. 3, 23, 11
6. 16; 360
7. 16; 13
8. (a) 4 h 44 min (b) 8 min 22 s (c) 2 h 48 min
9. 57°26′, 73°51′; (i) 32°35′ (ii) 68°20′ (iii) 77°11′

Exercise 93 (page 218)

2. 19.56 **3.** £51.20 **5.** 595; 34; 3003 **6.** 17; 1771

Exercise 94 (page 220)

1. (i) £47.86 (ii) £182.83 **2.** £29.03 **3.** 65p
4. £4.53 **5.** £58.75 **6.** 154.2 cm **7.** 168 cm^2
8. (i) 555.10 fr (ii) £55.04
9. (i) 142.22 dollars (ii) £189.09 **10.** £31.85

Revision Paper C1 (page 222)

1. (a) $9, \frac{1}{9}, 1, \frac{1}{3}$ (b) $2^4, 2^{-1}, 2^0, 2^{-3}$
(c) $a^{-5}, b^4, c^{-2}, d^{-1}$
(d) $7.3 \times 10^4, 8.4 \times 10^{-3}, 4 \times 10^6, 7 \times 10^{-6}$
2. (a) £1.32 (b) £220 **3.** (a) $\frac{1}{3}$ (b) 5
4. (a) $\bar{5}.4, \bar{3}.6, \overline{18}.5$ (b) (i) 0.612 (ii) 86.1
5. (i) 4 (ii) -3 (iii) -3 (iv) 2, 5 (v) 3, -4
6. (ii) (4, 3); 6 (iii) 15 (iv) 2; 63°26′
7. (a) Scalar in (i), vectors in the others (b) $\underline{a} = \underline{f}, \underline{d} = \underline{g}$
8. 30–40 thousand miles.

Revision Paper C2 (page 223)

1. (a) $0.36, 0.01, 900, 1\frac{9}{16}$ (b) $20, 0.2, 0.06, 1\frac{1}{2}$
(c) 8.19, 25.9, 0.819, 0.259
2. (a) (i) $\dfrac{4a + 7b}{28}$ (ii) $\dfrac{3d + 5c}{cd}$ (iii) $\dfrac{p + rn}{n^2}$
(b) (i) $\dfrac{e}{3} + \dfrac{f}{5}$ (ii) $\dfrac{x}{h} + \dfrac{u}{g}$ (iii) $\dfrac{1}{m} + \dfrac{7}{m^2}$
3. (i) (a) 61 kg (b) 79 kg (ii) $h = \frac{5}{3}(m + 35)$ (iii) 175 cm
4. (ii) (a), (b)
5. (i) 12.10 p.m. (ii) 10 km (iii) $\frac{1}{2}$ h (iv) 5 km/h (v) 4 km/h
6. (i) $a^2 + 10a + 25$ (ii) $b^2 - 16b + 64$ (iii) $c^2 + 4c - 21$
(iv) $d^2 - 6d + 8$
7. (i) 90° (ii) 38° (iii) 12.3 cm (iv) 24.6 cm

Revision Paper C3 (page 225)

1. (i) 25 litres (ii) 322 km
2. (a) $10^4, 10^{-2}, 10^{18}$ (b) $\frac{1}{4}, \frac{1}{27}, \frac{1}{16}$
(c) $1.2 \times 10^4, 6 \times 10^{-2}, 6.4 \times 10^4$
3. (a) (i) $a(a - 4)$ (ii) $(b + 2)(b - 2)$ (iii) $(c - 4)(c - 1)$
(iv) $(d - 4)(d + 1)$ (b) $-5, 1$
4. 1.30 m, 3.99 m
6. (a) $\begin{pmatrix} 2 \\ 4 \end{pmatrix}, \begin{pmatrix} 3 \\ -2 \end{pmatrix}, \begin{pmatrix} 5 \\ 2 \end{pmatrix}$ (b) The sum is the zero vector
7. (a) 38 475 m^2 and 42 775 m^2 (b) 5.27 km/h and 4.15 km/h
8. (i) 41.9 (ii) 6.45 (iii) 0.665 (iv) 0.444

Revision Paper C4 (page 226)

1. (a) $-1, -5, 3, 3, -28$ (b) $(A + B)(A - B)$ (i) 3200
(ii) 11.2 (iii) 11 (iv) 4
2. (i) False (ii) true (iii) false (iv) possible
3. (a) $\frac{1}{52}, \frac{1}{13}, \frac{1}{4}$ (b) $\frac{9}{20}$
4. 80 kg, 84 kg, 83 kg
5. (i) 32 degrees, 212 degrees (ii) 68, 14 (iii) $C = \frac{5}{9}(F - 32)$
(iv) 10, -15, 60
6. (a) (i) $3a^2 + 12a$ (ii) $b^2 - 2b - 35$ (iii) $c^2 - 6c + 9$
(b) (i) 0, 3 (ii) 5, -2 (iii) 5, -5
7. (i) 0.699 (ii) 23.5 (iii) 27.6 (iv) 104
8. 11.40; 10.50 and 11.23, $12\frac{1}{2}$ km and $20\frac{3}{4}$ km

Revision Paper C5 (page 228)

1. (i) 10 (ii) $3\frac{1}{2}, -2\frac{1}{2}$ (iii) -4, 3
2. (a) (i) 14.9 (ii) 6150 (iii) 0.412
(b) (i) 7.73 (ii) 20.6 (iii) 0.274
3. (i) (2, 1), (1, 3) (ii) $\widehat{ABC}$ and $\widehat{DEA}$ (iii) AB = BC = DE = EA
(iv) 8
4. (a) (i) $\begin{pmatrix} 3 \\ 5 \end{pmatrix}$ (ii) $\begin{pmatrix} 10 \\ 1 \end{pmatrix}$ (iii) $\begin{pmatrix} 3 \\ 1 \end{pmatrix}$
5. (i) After 15 s and 80 s (ii) 0.8 m/s^2 (iii) 1.2 m/s^2 (iv) 1560 m
6. (i) 0.0791 (ii) 12.6 (iii) 0.503 (iv) 0.958
7. 106 cm, 14°45′; 176 cm, 25
8. (a) (i) 3, 5 (ii) 7, 1 (iii) To find the quotient and remainder when a number is divided by 7